Trace Analysis
with Microcolumn
Liquid Chromatography

CHROMATOGRAPHIC SCIENCE SERIES

A Series of Monographs

Editor: JACK CAZES
Sanki Laboratories, Inc.
Mount Laurel, New Jersey

1. Dynamics of Chromatography, *J. Calvin Giddings*
2. Gas Chromatographic Analysis of Drugs and Pesticides, *Benjamin J. Gudzinowicz*
3. Principles of Adsorption Chromatography: The Separation of Nonionic Organic Compounds, *Lloyd R. Snyder*
4. Multicomponent Chromatography: Theory of Interference, *Friedrich Helfferich and Gerhard Klein*
5. Quantitative Analysis by Gas Chromatography, *Josef Novák*
6. High-Speed Liquid Chromatography, *Peter M. Rajcsanyi and Elizabeth Rajcsanyi*
7. Fundamentals of Integrated GC-MS (in three parts), *Benjamin J. Gudzinowicz, Michael J. Gudzinowicz, and Horace F. Martin*
8. Liquid Chromatography of Polymers and Related Materials, *Jack Cazes*
9. GLC and HPLC Determination of Therapeutic Agents (in three parts), *Part 1 edited by Kiyoshi Tsuji and Walter Morozowich, Parts 2 and 3 edited by Kiyoshi Tsuji*
10. Biological/Biomedical Applications of Liquid Chromatography, *edited by Gerald L. Hawk*
11. Chromatography in Petroleum Analysis, *edited by Klaus H. Altgelt and T. H. Gouw*
12. Biological/Biomedical Applications of Liquid Chromatography II, *edited by Gerald L. Hawk*
13. Liquid Chromatography of Polymers and Related Materials II, *edited by Jack Cazes and Xavier Delamare*
14. Introduction to Analytical Gas Chromatography: History, Principles, and Practice, *John A. Perry*
15. Applications of Glass Capillary Gas Chromatography, *edited by Walter G. Jennings*
16. Steroid Analysis by HPLC: Recent Applications, *edited by Marie P. Kautsky*
17. Thin-Layer Chromatography: Techniques and Applications, *Bernard Fried and Joseph Sherma*
18. Biological/Biomedical Applications of Liquid Chromatography III, *edited by Gerald L. Hawk*
19. Liquid Chromatography of Polymers and Related Materials III, *edited by Jack Cazes*
20. Biological/Biomedical Applications of Liquid Chromatography, *edited by Gerald L. Hawk*

21. Chromatographic Separation and Extraction with Foamed Plastics and Rubbers, *G. J. Moody and J. D. R. Thomas*
22. Analytical Pyrolysis: A Comprehensive Guide, *William J. Irwin*
23. Liquid Chromatography Detectors, *edited by Thomas M. Vickrey*
24. High-Performance Liquid Chromatography in Forensic Chemistry, *edited by Ira S. Lurie and John D. Wittwer, Jr.*
25. Steric Exclusion Liquid Chromatography of Polymers, *edited by Josef Janča*
26. HPLC Analysis of Biological Compounds: A Laboratory Guide, *William S. Hancock and James T. Sparrow*
27. Affinity Chromatography: Template Chromatography of Nucleic Acids and Proteins, *Herbert Schott*
28. HPLC in Nucleic Acid Research: Methods and Applications, *edited by Phyllis R. Brown*
29. Pyrolysis and GC in Polymer Analysis, *edited by S. A. Liebman and E. J. Levy*
30. Modern Chromatographic Analysis of the Vitamins, *edited by André P. De Leenheer, Willy E. Lambert, and Marcel G. M. De Ruyter*
31. Ion-Pair Chromatography, *edited by Milton T. W. Hearn*
32. Therapeutic Drug Monitoring and Toxicology by Liquid Chromatography, *edited by Steven H. Y. Wong*
33. Affinity Chromatography: Practical and Theoretical Aspects, *Peter Mohr and Klaus Pommerening*
34. Reaction Detection in Liquid Chromatography, *edited by Ira S. Krull*
35. Thin-Layer Chromatography: Techniques and Applications. Second Edition, Revised and Expanded, *Bernard Fried and Joseph Sherma*
36. Quantitative Thin-Layer Chromatography and Its Industrial Applications, *edited by Laszlo R. Treiber*
37. Ion Chromatography, *edited by James G. Tarter*
38. Chromatographic Theory and Basic Principles, *edited by Jan Åke Jönsson*
39. Field-Flow Fractionation: Analysis of Macromolecules and Particles, *Josef Janča*
40. Chromatographic Chiral Separations, *edited by Morris Zief and Laura J. Crane*
41. Quantitative Analysis by Gas Chromatography, Second Edition, Revised and Expanded, *Josef Novák*
42. Flow Perturbation Gas Chromatography, *N. A. Katsanos*
43. Ion-Exchange Chromatography of Proteins, *Shuichi Yamamoto, Kazuhiro Nakanishi, and Ryuichi Matsuno*
44. Countercurrent Chromatography: Theory and Practice, *edited by N. Bhushan Mandava and Yoichiro Ito*
45. Microbore Column Chromatography: A Unified Approach to Chromatography, *edited by Frank J. Yang*
46. Preparative-Scale Chromatography, *edited by Eli Grushka*
47. Packings and Stationary Phases in Chromatographic Techniques, *edited by Klaus K. Unger*
48. Detection-Oriented Derivatization Techniques in Liquid Chromatography, *edited by Henk Lingeman and Willy J. M. Underberg*

49. Chromatographic Analysis of Pharmaceuticals, *edited by John A. Adamovics*
50. Multidimensional Chromatography: Techniques Applications, *edited by Hernan Cortes*
51. HPLC of Biological Macromolecules: Methods and Applications, *edited by Karen M. Gooding and Fred E. Regnier*
52. Modern Thin-Layer Chromatography, *edited by Nelu Grinberg*
53. Chromatographic Analysis of Alkaloids, *Milan Popl, Jan Fähnrich, and Vlastimil Tatar*
54. HPLC in Clinical Chemistry, *I. N. Papadoyannis*
55. Handbook of Thin-Layer Chromatography, *edited by Joseph Sherma and Bernard Fried*
56. Gas–Liquid–Solid Chromatography, *V. G. Berezkin*
57. Complexation Chromatography, *edited by D. Cagniant*
58. Liquid Chromatography–Mass Spectrometry, *W. M. A. Niessen and J. van der Greef*
59. Trace Analysis with Microcolumn Liquid Chromatography, *Miloš Krejčí*

Additional Volumes in Preparation

Modern Chromatographic Analysis of Vitamins, Second Edition, *André P. De Leenheer, Willy E. Lambert, and H. Nelis*

Preparative and Production Scale Chromatographic Methods and Applications, *edited by G. Ganetsos and P. E. Barker*

Trace Analysis with Microcolumn Liquid Chromatography

Miloš Krejčí

Institute of Analytical Chemistry
Czechoslovak Academy of Sciences
Brno, Czechoslovakia

Marcel Dekker, Inc.   New York • Basel • Hong Kong

Library of Congress Cataloging-in-Publication Data

Krejčí, Miloš.
   Trace analysis with microcolumn liquid chromatography / Miloš
Krejčí.
      p.   cm. – (Chromatographic science ; v. 59)
      Includes bibliographical references and index.
      ISBN 0-8247-8641-6 (alk. paper)
      1. Liquid chromatography.   2. Trace analysis.   I. Title.
   II. Title: Microcolumn liquid chromatography.   III. Series.
   QD79.C454K74   1992
   543'.0894–dc20   92-4159
                                                                   CIP

This book is printed on acid-free paper.

MARCEL DEKKER, INC.
270 Madison Avenue, New York, New York 10016

Current printing (last digit):
10 9 8 7 6 5 4 3 2 1

PRINTED IN THE UNITED STATES OF AMERICA

Preface

The sharp rise in popularity and the recent development of apparatus for microbore column liquid chromatography for trace analysis make this volume timely. The theoretical discussion establishes the basis of liquid chromatography with regard to the role played by column inner diameter. The optimal column parameters of packed microbore and open tubular column liquid chromatography are derived. In the discussion of packed small-bore columns, great attention is paid to instrumentation, column preparation, and the demands on detectors and sampling devices of the solute in microbore column liquid chromatography. Techniques of work with sorbents of small-particle diameter (3 μm) are also presented. Further, the theoretical basis of trace analysis of samples of both very-low-mass and very-low-solute concentrations is surveyed.

A detailed description of technical elements of a microbore column liquid chromatograph suitable for use in trace analysis is given. Different possibilities with enrichment columns in microbore column liquid chromatography are advanced. Examples of analyses, especially from the spheres of biochemistry, pharmacology, and environmental analytical chemistry, are provided. The apparatus components for work with open-tubular capillary columns are examined and the preparation of both fused silica and glass capillary columns is described here. The closing chapter reviews combinations of microbore column liquid chromatography with identification spectral procedures, especially these combinations: liquid chromatography–mass spectrometry and liquid chromatography–infrared spectrometry with Fourier's transformation.

Trace analysis is the focus of interest in analytical chemistry today. More efficient separations of analytes from complex natural and artificial mixtures with low concentrations and masses of the analytes are demanded. Ever-developing microcolumn liquid chromatography is used for both concentration and mass trace analysis.

Attention is devoted to some phenomena specific to the use of small-bore columns (Marangoni's effect in open-tube capillary columns) and to the phenomena occurring in liquid chromatography and affecting its quantitative results in general. Knowledge of these phenomena will enable the readers to orient themselves to more sophisticated chromatographic experiments.

The majority of companies manufacturing laboratory equipment offer microcolumn versions of liquid chromatographs. Readers will come to understand underlying theoretical principles while extending their familiarity with applications of microcolumn liquid chromatography in practice.

MILOŠ KREJČÍ

Contents

Preface   **iii**

1 Trace Analysis   **1**
  1.1 Significance of Trace Analysis and Its Basic Types   1
  1.2 Role of Chromatography   4
       References   7

2 Miniaturization   **8**
  2.1 Advantages   8
  2.2 Brief Summary of Theoretical Principles   12
  2.3 Terminology   15
  2.4 History of the Development of Liquid Chromatography
      with Small-Bore Columns   17
       References   21

3 Microcolumns   **26**
  3.1 Theoretical Bases   26
  3.2 Apparatus   38
       References   90

4 Trace Analysis by Microcolumn Liquid Chromatography   94
   4.1   Trace Concentration of the Analyte   94
   4.2   Peak Focusing Techniques   97
   4.3   Enrichment Columns   114
        References   120

5 Capillary Columns   122
   5.1   Pumps and Injection Systems   124
   5.2   Capillary Columns   125
   5.3   Detectors   152
        References   162

6 Examples of Analysis   165
   6.1   Biological Samples   166
   6.2   Pharmaceutical Samples   174
   6.3   Inorganic Analysis   178
        References   185

**7 Combination of Microcolumn Liquid Chromatography
   with Spectral Identification Methods   187**
   7.1   Combination with Mass Spectrometry   188
   7.2   Combination with Infrared Spectrometry   193
        References   198

Index   201

1
Trace Analysis

1.1. SIGNIFICANCE OF TRACE ANALYSIS AND ITS BASIC TYPES

The development of science and technology, particularly in the 1950s and later, led to the immense growth of the significance of analytical chemistry. Standards of quality of materials, as well as the necessity of exact control of technological procedures in the production of these materials, emphasized the importance of industrial analysts. The development of molecular aspects of some sciences, including medicine, biology, and genetics in particular, places ever-increasing demands on analysts in scientific research. Legal measures concerned with the chemical aspects of the environment, the hygiene of foodstuffs, drug abuse, and other areas elevated the role of public analysts. Higher standards are set for both qualitative and quantitative analysis. Mixtures of substances subjected to analyses often consist of as many as several hundreds up to several thousands of individual components with very low concentrations of significant components.

At the present time most of the activities of analysts are aimed at problems associated with determinations of small masses or low concentrations of analytes. In some instances, determination of low concentrations in a small mass or in a small volume of sample usually gives rise to the most exacting problems. An outline of relevant volumes (masses) of samples and relevant concentrations of analytes in a sample is presented in Figure 1.1. From the outline a nomenclature proposal for the field of trace analysis can be derived. The components present in the sample at concentrations (g/ml, g/g, mol/ml, μmol/g, and so on) greater than 10^{-2}, that is, components in an amount greater than 1%, are major components. Minor components are those in concentrations of 10^{-4} (hundredths of a percent). Trace components or trace analytes occur at concentrations in the range from 10^{-4} to 10^{-12}. Even lower concentrations are designated micro- or nanotrace components. This nomenclature has its origins in the IUPAC proposal [1]; however, it cannot be considered obligatory or restrictive.

The attention and activity of analysts have obviously been associated with trace analysis. In random perusal of any volume of *Chemical Abstracts* we discover that works dealing with liquid chromatography, for instance, also concern the field of trace analysis more than 50% of the time. It is also of interest that only several tenths of a percent of these publications include the term "trace analysis" in the title. We can thus conclude that the majority of authors consider the term "trace analysis" to be synonymous with the term "analysis."

In any analysis it is necessary to determine the quantity of analytes in samples. In the outline in Figure 1.1 the quantity of the analyte is represented by the product of corresponding values on the x axis, that is, the volume or mass of the sample, and those on the y axis, that is, the analyte concentration. Knowing the quantity of the analyte and its volume (mass), we can calculate analyte concentration in the sample. The least value of the product of both these magnitudes gives the minimum analyte amount that can be determined. Based on this principle, trace analysis can be classified into two categories: (1) mass trace analysis and (2) concentration trace analysis. Mass trace analysis requires analysis of small amounts of sample. As a consequence, the minimum determinable concentration acquires a higher value than when the amount of the sample under analysis is sufficiently large. Such analyses are frequently biomedical, such as analyses of biologically

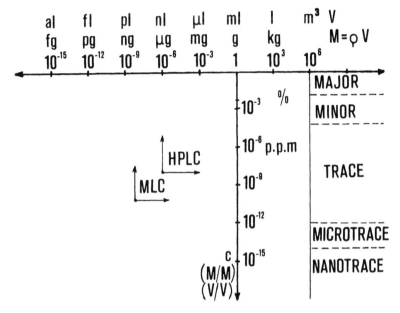

Figure 1.1 Terminology and units in trace analysis; V = volume; M = mass; ρ = density.

active elements (pheromones). In trace analyses of this type miniaturized modes of chromatography play an important role.

For concentration trace analysis, a larger volume (mass) of sample is usually available. Under these circumstances concentrations of 10^{-11} and lower can be determined. In the majority of this type special techniques for sample enrichment, such as extraction and concentration, are required. Environmental analysis, particularly analyses of water and the atmosphere, is a typical example.

In evaluating the development of trace analysis, we cannot ignore that the numbers of single-purpose analyses keep increasing, not only in production technology or environmental control and quality of foodstuffs, but also diverse research activities in geology, epidemiology, and other fields. The same trend can also be noted in the development of many scientific fields in which only large series of analyses permit the statistical processing of observed phenomena. This trend leads on the one hand to the development and production of automatically operating analyzers and, on the other, to fully automated laboratory instruments. In trace analysis those procedures that make it possible to fully utilize automatic laboratory devices are therefore preferred. In chromatography both enrichment processes in special enrichment columns and direct analyte enrichment in the chromatographic columns are involved.

1.2. ROLE OF CHROMATOGRAPHY

Trace concentrations of analytes in samples are accompanied by a considerably complex composition of analyzed mixtures. The smaller the analyte concentration to be determined, the larger the number of components in the mixture that can be expected. Under such circumstances sufficient selectivity of the analysis cannot be reached without separating the individual components of the mixture or, at least, without separating their groups. Chromatography, in its capacity as an analytical separation method that combines high separation efficiency with the possibility of determining extremely low concentrations or masses of analytes, is irreplaceable in trace analysis.

The principle of the separation of analytes A to M can be expressed in a simplified way [2] as

$$[A + B + C + \cdots + M] \rightarrow [A] + [B] + [C] + \cdots + [M] \quad (1.1)$$

where A, B, C, . . ., are the amounts (moles or mass) of individual components in a sample. Equation (1.1) can be rewritten using sample volume V_S and concentrations (mol/L or g/L) of individual components in the sample, c_{SA}, c_{SB}, and so on, into the form

$$m_S = V_S [c_{SA} + c_{SB} + c_{SC} + \cdots + c_{SN}] \rightarrow [V_{iA} \, \hat{c}_{iA}] + [V_{iB} \, \hat{c}_{iB}] + [V_{iC} \, \hat{c}_{iC}] + \cdots + [V_{iN} \, \hat{c}_{iN}] \quad (1.2)$$

where m_S is the total amount of the sample and V_i is the mobile-phase volume in which the component is eluted at an average concentration $\hat{c}_i$. It holds that

$$\sum_A^M c_{SI} = 1 \quad \text{or} \quad \sum_A^M V_{iI}\, \hat{c}_{iI} = m_S \tag{1.3}$$

The mass of the selected analyte (e.g., A) is usually described with sufficient precision as

$$A = (c_{max}\sigma_{vol}K) \tag{1.4}$$

where c_{max} is the concentration of analyte A at the outlet of the chromatographic column, σ_{vol} is the volume standard deviation of that peak, and K is the proportionality constant. The total sample mass can be then expressed as

$$m_s = \left(\frac{Kc_{max}V_R}{\sqrt{n}}\right)_A + \left(\frac{Kc_{max}V_R}{\sqrt{n}}\right)_B \\ + \cdots + \left(\frac{Kc_{max}V_R}{\sqrt{n}}\right)_M \tag{1.5}$$

where V_R is the retention volume and n is the number of theoretical plates. As is shown later, the retention volume V_R decreases as the column diameter decreases, the analyte concentration at the column outlet increases as the column diameter decreases, and the number of theoretical plates is independent of the diameter of the packed column d_c. Hence it holds that

$$V_R = f(d_c)$$
$$c_{max} = f(d_c) \tag{1.6}$$
$$n \neq f(d_c)$$

Packed columns play an important part in trace analysis. Open-tube capillaries, whose efficiency (the number of theoretical plates) depends on their inner diameters, have not yet gained much importance. Considering trace analysis, we start from the fact that the minimum detectable response R_{det} is a function of the analyte concentration in the peak maximum at the column outlet:

$$R_{det} = f(c_{max}) \tag{1.7}$$

The response must exceed the detector noise at least twofold.

For concentration trace analysis we can write

$$[A] + [B] \gg \sum_c^M I \tag{1.8}$$

This means that the components representing the sample matrix (in our example components A and B in Equation (1.8) are concerned) are present in the sample in an amount substantially greater than that of minor and trace compounds I. In practice, matrix components represent, for example, water, blood plasma, or monomer in the analysis of polymerization additives. In most analyses these components are not determined. In chromatographic analysis, however, they represent problems associated with the separation of trace components from the matrix or with sufficient selectivity of the chromatographic detection system. Converting Equation (1.8) into dimensions of concentrations, we can write

$$c_{SA} + c_{SB} \gg \sum_c^M c_S I \tag{1.9}$$

with simultaneous validity of

$$c_{SA} + c_{SB} \rightarrow 1 \qquad c_{SA} + c_{SB} < 1 \tag{1.10}$$

In view of the condition for the detection limit in Equation (1.7) and in view of Equation (1.4), a certain minimum concentration at the column outlet obviously exists that can be determined, and simultaneously, a certain minimum amount of analyte in the same is also required for this determination. Mass trace analysis is discussed here in agreement with this conclusion. It follows from these relationships and Figure 1.1 that in concentration analysis extremely low analyte concentrations can be determined if the sample is sufficiently large. In mass trace analysis the concentrations that can be determined are higher because of the limited sample size available.

Liquid chromatography plays an important role in trace analysis of both types. An indisputable advantage of microcolumn liquid chromatography is its enhanced sensitivity in mass trace analysis. For concentration trace analysis in liquid chromatography conventional columns can also be applied successfully. The lesser amounts of samples required for achievement of the same detection limits, however, suggest that microcolumn liquid chromatography also has advantages in this respect. Utilization of enrichment columns as well as enrichment at the microcolumn inlet make it possible to classify microcolumn liquid chromatography as one of the most efficient current analytic procedures.

REFERENCES

1. Irving P. M., Freiser W., West F. V.: IUPAC Compendium of Analytical Nomenclature, Pergamon Press, Oxford, 1978.
2. Giddings J. C.: J. Chromatogr. 395, 19 (1987).

2
Miniaturization

2.1 ADVANTAGES

The rapid development of gas chromatography in the 1950s and particularly in the 1960s led to a qualitative upheaval in the development of analytic chemistry. The possibilities for fast and perfect separations of components from complex mixtures of substances greatly increased the informational value of analytic results. Establishment of the theory of gas chromatography together with a shift in the interest of the scientific public to the field of high-molecular-weight nonvolatile organic and inorganic substances led in the 1960s to a renewal of column chromatography with a liquid mobile phase, that is, high-performance liquid chromatography. This new development gave rise to extension of the gas chromatography theory to the domain of liquid chromatography in terms of both sorption equilibria and system selectivity, and dynamics [1–3] of nonequilibrium chromatographic systems. Refinement in instrumentation [4,5] proceeded together with this development so that in the mid-1970s it was possible to work on the level of theoretically obtainable separations using commercially available instruments.

Over the course of the last decade two important trends can be traced in the development of liquid chromatography. The application of integrated electronics, microprocessor techniques in particular, led to the almost complete automation of chromatography. The other significant trend was reflected in the miniaturization of liquid chromatography.

The present development of the technology is characterized by the tendency toward miniaturization that we encounter not only in the field of electronics but also in the field of analytic instruments. Miniaturization in its present stage of development is associated with three fundamental conditions:

1. Miniaturized systems must maintain or improve the properties of conventional systems.
2. The energy supply for the operation of these systems must be reduced.
3. Material savings are usually fulfilled automatically even though quantitative saving is sometimes accompanied by more stringent demands on the quality of the construction material.

Material savings, that is, reduced consumption of the mobile and stationary phases, are usually emphasized in connection with the advantages of microcolumn liquid chromatography. It is really possible to use rare and as a result also expensive liquids as mobile phases and simultaneously also suppress the costs of the sorbents used in microcolumns. These advantages are not so important, however, as to hold out much hope of a wider application of miniaturization in chromatographic practice.

However, the importance of miniaturization in liquid chromatography is evident, first, in new and better possibilities in the field of analytic separation and in the field of trace analysis by liquid chromatography. The significance of miniaturization may be summarized as follows.

It follows from liquid chromatography theory that all separations that can be performed by conventional methods can also be effected in miniaturized systems. Moreover, miniaturization offers the possibility of reaching greater efficiencies, 10^5–10^7 theoretical plates. The reason may be in improved heat transfer from microcolumns and consequently also suppression of interfering effects of frictional heat in the application of higher pressures and the possibility of constructing long columns. Higher pressures in microcolumn chromatography permit faster analysis than with chromatographs equipped with conventional columns.

In microcolumn liquid chromatography the same minimum detectable concentrations can be reached as in conventional chromatography. Samples with a lower minimum detectable mass than in conventional columns can be analyzed in microcolumn systems. Every chromatographic separation involves a dilution procedure. The ratio of the analyte concentration in the sample c_S to the analyte concentration in the peak maximum at the column outlet is dependent on the chromatographic column diameter, however. The ratio c_S/c_{max} decreases as the column diameter is reduced. As a result, microcolumn chromatography makes it possible to design enrichment systems more easily and more effectively. Under suitable conditions the enrichment can be obtained directly in microcolumns, where $c_S < c_{max}$, and consequently, it is possible to work with high concentration sensitivity. These results can be achieved with substantially smaller samples than are necessary in columns of larger diameter.

Decreased dilution of analytes in chromatographic separations, that is, a decreasing c_S/c_{max} ratio, facilitates liquid chromatographic coupling with spectral methods. The advantages are particularly obvious if the spectrometric response has the character of a mass response (for instance in mass spectrometry). Small retention volumes of solutes in microcolumn chromatography do not set such stringent standards for providing a vacuum in the mass spectrometer. Increased solute concentration at the column outlet is also positive, however, in other spectrophotometric techniques, such as spectrophotometry in the ultraviolet and visible regions of spectrum (UV/VIS) with a photodiode array, infrared (IR) spectrophotometry, and nuclear magnetic resonance (NMR).

Miniaturization makes it possible to design qualitatively new instruments. For instance, a chromatograph with the combined functions of sample injection and mobile-phase transport creates conditions for the construction of fully automated chromatographic analyzers. It permits the preparation of special mobile phases for any analysis. Changing the chromatographic system does not require washing dead spaces with negligible dimensions. Once selected, new mobile phases can be used immediately. New systems of gradient techniques are applicable to miniaturized instruments, thereby offering new possibilities for the field of trace analysis. Coupling such a chromatograph with spectrophotometers aims, in view of the very small retention volumes and relatively high solute concentrations at the column outlet, toward the creation of a complex analytic instrument whose operation is entirely controlled by computer. Instruments of this type can be utilized as sensors for the robotization of analytic factories.

The very small retention volumes necessary for the separation of the given components while maintaining the capacity ratios of the components permit utilization of mobile phases composed of rare and expensive chemicals. In view of the small volumes due to liquid compression, work with liquid chromatographs becomes safer.

Any miniaturization brings new technical requirements for instrumentation. Although the principles of various elements do not differ from those of most classic instruments, some parameters must be different. The technique for the preparation of miniaturized packed columns in liquid chromatography has substantially been solved. As far as capillary liquid chromatography columns are concerned, some problems have remained unsolved. Substantially higher demands than in chromatography with conventional columns are placed on the minimization of extracolumn spaces. At the present time attention is concentrated upon the question of miniaturization of concentration-dependent detector cells and injection techniques. Better automation of miniaturized liquid chromatographs remains an open problem.

The prospects for the further development of miniaturization in liquid chromatography follow from theoretical analyses. As in other fields, miniaturization will penetrate the industrial production of liquid chromatographs when new analytic possibilities for miniaturized systems are available and in wide demand. Under these circumstances the economic advantages need not be conditional for the spread of miniaturized systems. Unless the advantages of microcolumn liquid chromatography are evident and used in mass production, miniaturized systems will hardly be competitive with today's conventional instruments, which are reliable in operation within the limits of theoretically obtainable parameters. However, the existing trend suggests that the number of chromatograph models with microcolumns available on the world market keeps growing. Microchromatography thus ranks among the methods used in analytic laboratories despite not having reached all theoretically forecasted advantages of miniaturized systems in all instances.

The development of miniaturization seems to be completing the process started by Golay [6], who introduced capillary columns to gas chromatography. The parameters theoretically forecast for liquid chromatography [2] are being reached, and the possibilities are being slowly approached to analyze volumes so small that it is possible to study the contents of the single cell of an organism, as once the founder of modern chromatography, Martin [7], predicted.

2.2 BRIEF SUMMARY OF THEORETICAL PRINCIPLES

It is appropriate for readers who are not familiar with the theoretical principles of chromatography to present here a brief summary of some basic terms and important relationships and their significance. It will make orientation in the ensuing text easier.

The theory of the method is important for practical analysts since it allows them to determine or estimate suitable experimental conditions. The theory considers three characteristics that are easy to read from the chromatogram:

1. Position of the peak in the chromatogram
2. Shape of the peak, most frequently its width at a selected fraction of its height
3. Size of the peak, that is, its surface or its height provided that the peak shape is symmetrical

The first characteristic, the peak position in the chromatogram, is most frequently characterized by retention time t_R:

$$t_R = t_M(1 + k) \tag{2.1}$$

where the dead time t_M corresponds to the elution time of an unsorbed peak:

$$t_M = \frac{L}{u} \tag{2.2}$$

where L is the column length and u is the linear velocity of the mobile phase.

The capacity ratio k characterizes the extent of analyte retention in the chromatographic column and is expressed by

$$k = \frac{t_R - t_M}{t_M} - \frac{t_R'}{t_M} - \frac{K_D V_s}{V_m} \tag{2.3}$$

The capacity ratio is thus proportional to the distribution constant K_D and to the ratio of the stationary-phase volume V_s to the mobile-phase volume V_m. The product of the retention time and the mobile-phase flow rate F_m determines the retention volume V_R:

$$V_R = t_R F_m = V_M(1 + k) = \frac{\pi}{4} d_c^2 L \varepsilon_u(1 + k) \tag{2.4}$$

where V_M is the dead volume, d_c is the column diameter, and ε_u is the interstitial porosity of the sorbent bed through which the mobile phase flows. The reduced retention time t_R' or the reduced retention volume V_R' is

$$V_R' = t_R' F_m = (t_R - t_M)F_m = V_R - V_M \tag{2.5}$$

The efficiency of the chromatographic column is usually expressed in terms of the number of theoretical plates n:

$$n = \frac{V_R^2}{\sigma_V^2} = \frac{L}{H} = \frac{L}{hd_p} = \frac{\lambda}{h} \tag{2.6}$$

where σ_V^2 is the square of the peak standard deviation (variance) expressed in volumetric units, d_p is the mean diameter of particles of the sorbent, h is the reduced height equivalent to a theoretical plate, and λ is the reduced column length ($\lambda = L/d_p$).

It is obvious from Equation (2.6) that the height equivalent to a theoretical plate H represents the measure of the broadening of the chromatographic peak per unit of column length. With respect to the fact that the diffusion path of the solute in the column is determined either by the sorbent particle diameter d_p (in packed columns) or by the inner diameter of the open-tube capillary column d_c, so-called reduced quantities are used to advantage in chromatography. The height equivalent to a theoretical plate H is dependent on the mobile-phase flow rate. Characterizing flow rate, we use volumetric units,

$$F_m = \frac{\pi}{4} u d_c^2 \, \varepsilon_u \tag{2.7}$$

linear velocities of the mobile phase u,

$$u = \frac{L}{t_M} \tag{2.8}$$

or the reduced velocities v,

$$v = \frac{u d_p}{D_m} \tag{2.9}$$

where D_m is the diffusion coefficient.

The quantitative content of the analyte in the mixture is proportional to the size of the corresponding peak in the chromatogram. Peak integration is today in the majority of instances a question of automatic evaluation of chromatograms. As long as the peak concentration profile can be described by the Gaussian function, not only integrated peak area, but also the product of the peak height and its width at a defined fraction of the peak height, as well as simple peak height, is representative. In connection with microcolumn chromatography, the question of the solute concentration in the peak maximum is important. The value of c_{max} at the column outlet is expressed as

$$c_{max} = c_S V_S \frac{\sqrt{n}}{V_R \sqrt{2\pi}} \tag{2.10}$$

where c_S is the solute concentration in the sample and V_S is the sample volume.

2.3 TERMINOLOGY

The entry of miniaturization into liquid chromatography gave rise to nomenclatural problems. The term "microcolumn liquid chromatography" by itself expresses only that the column diameter is reduced. The development of theory with subsequent intensive development of instrumentation, however, required a more detailed nomenclatural distinction for chromatography in columns with small diameters.

The basic classification of miniaturized columns in liquid chromatography follows from the principle of peak dispersion in the chromatographic column. In columns of one type the diffusion path of the solute is determined by the size of particles of the sorbent packed in the column; in columns of the other type it is proportional to the column diameter. On this basis the classification of small-bore columns was proposed [8]:

> Type A, microcolumns, packed columns of small diameter, represent substantially conventional analytic columns of reduced diameter. Peak dispersion in the column is determined by the size of the sorbent particles.
>
> Type B, packed capillary columns, include irregular packed columns in which the diameter of the sorbent particles approaches the column diameter. This represents columns of a transition type.
>
> Type C, capillary columns, are open tubular capillaries in which the diffusion path of the solute is a function of the column diameter. The proposed classification of miniaturized columns in liquid chromatography starts from the reduced values of column diameter ζ

$$\zeta = \frac{d_c}{d_p} \tag{2.11}$$

> where d_c is the inner column diameter and d_p the mean sorbent particle diameter.

The reduced column diameter can also be used for a more detailed column classification [9] used in present liquid chromatography (Table

Table 2.1. Proposal for Column Nomenclature Based on Reduced Column Diameter

Reduced Column Diameter	Column Diameter d_c	Type	Columns
Capillary columns	5–10 μm	0	Open-tube capillary columns for HPLC
Columns with small diameter			
1–100	30–500 μm	I	Packed capillary cols.
100–600		II	Microcolumns
100–300	>0.5–1 mm	IIA	
300–600	>1–2 mm	IIB	
Conventional columns, 600–2,000	3–6 mm	III	Conventional analytical columns
Preparative columns, above 2,000	>6 mm	IV	Preparative columns

Source: From Reference 9.

2.1). However, the reduced diameter is not defined for capillary columns since sorbent is not used. The reduced diameter ζ obviously acquires roughly the values from 1 to 200. The author of the proposal correctly assumes that the values of d_p are in the interval from 3 to 10 μm. Capillary columns and microcolumns are included in the first two groups in Table 2.1 (according to the principle of the diffusion path), and here they are further classified into five categories by the size of the reduced column diameters.

This complicated classification is particularly justified for the description and evaluation of the properties of instrumental elements in front of the column (a device for the generation of the mobile-phase gradient, the sample injector) or connected behind the column (the detector). From the viewpoint of chromatographic separation the classification of small-bore columns into two groups can be considered entirely satisfactory.

This book uses original terminology. Columns of inner diameters of the order of tens to hundreds of micrometers (sometimes up to 1 mm) packed with a sorbent whose particle size only exceptionally exceeds 10 μm are called microcolumns. Capillaries of inner diameter from 5 to 100 μm are called capillary columns.

This work further required the use of several terms that are not very common in the chromatographic literature. The retention strength of the chromatographic system is expressed quantitatively by the magnitude of the capacity ratio of a given solute: it is larger the higher the capacity ratio of the solute. The elution strength of the mobile phase is a constituent of the system retention strength, and its increase leads to a reduction in the system retention strength. An increase in the retention strength of the stationary phase results in an increase in the retention strength of the system.

The term "injection-generated gradient" designates such a technique for the generation of the system retention strength in which a modifier that affects the elution strength of the mobile phase or the retention strength of the stationary phase is added to the analyzed sample. With respect to the character of modifier migration through the chromatographic column, both phases are influenced by the modifier to the extent given by the sorption isotherm of the modifier in the given chromatographic system.

Sample enrichment by the analyte leads to an increase in its concentration in the sample. *Peak focus* is the name of the technique leading to a reduction in the volume of the effluent in which the given analyte is eluted out of the column. *Dispersion* is a reversed process.

2.4 HISTORY OF THE DEVELOPMENT OF LIQUID CHROMATOGRAPHY WITH SMALL-BORE COLUMNS

In the late 1960s and in the early 1970s the first works dealing with small-diameter columns appeared in the literature. In this period increased attention was devoted to the velocity of the mass transfer between the phases. The first works [10,11] were therefore associated with the use of pellicular particles and solved the problems of ion exchange in liquid chromatography.

The first works showing the possible applications of capillary columns in liquid chromatography [12] originated in this period. As late as 1978, 15 works appeared that dealt with the technique of capillary columns [13–18] or microcolumns [19–22], and this meant the actual beginning of miniaturization in liquid chromatography. This period was dominated by efforts to cope technically with the miniaturization of instruments and to suppress peak dispersion in extracolumnar chromatographic spaces [23], as well as to apply new detection principles [24,25] to microcolumn and capillary liquid chromatography [18].

After being classified systematically, the results obtained, together with the theoretical background [8,26] created, became the starting point for the further radical development of miniaturization in liquid chromatography.

Micropacked columns were developed from the viewpoint of their design in two directions. One was represented by columns of inner diameter of approximately 1 mm (e.g., Refs. 27 and 28). This direction is especially preferred by Western European and American manufacturers of chromatographs. A column diameter of 1 mm (in some instances even columns of inner diameter of 2 mm are included in the micropacked column category) sets substantially lower standards for the miniaturization of further elements of chromatography, such as detectors, injectors, and extracolumnar connectors in particular.

Columns in the other design direction have an inner diameter of 0.7–0.1 mm [29–42]. Small column diameters may set higher standards for their miniaturization; on the other hand, they offer the possibility of instrumentation innovations of a higher order. Small column diameters make it possible to use high pressures of the mobile phase and, as a result of a good removal of frictional heat, a high efficiency characterized by the number of theoretical plates is reached. The use of short columns [43–47] enables us to achieve the very short times necessary for chromatographic analysis. The speed of the analysis and very small volumes of the mobile phase required for an analysis, together with an excellent theoretical separation efficiency, thus establish a good basis for innovations of liquid chromatographs of higher orders.

From the viewpoint of function, attention is today paid to columns that reach high numbers of theoretical plates ($n > 10^4$) and to short columns, permitting very fast analyses. For the former type of columns quartz capillaries with an inner diameter of 0.2 mm and length of several tens to several hundred centimeters are usually selected [48,49]. The majority of current problems in practical analytic chemistry can be solved in short columns of the latter type (3–15 cm in length). These columns have no more than 10^3 theoretical plates [50]. The successful use of these columns [40,43–47,51] is dependent on the use of injectors and detectors that satisfy demands for the miniaturization of liquid chromatography.

Capillary liquid chromatography makes it possible to perform analyses with high separation efficiency in a short time and sets lower standards for the pressure gradient than microcolumn chromatography. It has not yet reached such a boom as microcolumn chromatography, and its effective use can be expected only when more than 10^6 theoretical plates are required [8]. Although such efficiencies have been obtained in practice [52,53], technical difficulties of such significance occurred that not a single manufacturer has been found who would produce capillary liquid chromatographs commercially. It seems at present that despite the growth of demands for the separation of components with very close physicochemical properties, most of the problems in analytic chemistry can be solved by combining selective and efficient elements of the chromatographic system with the use of lower efficiency columns.

Since the values of diffusion coefficients in the liquid phase are lower than those acquired by diffusion coefficients in the gaseous phase, columns of very small inner diameters must be chosen for liquid open-tube capillary chromatography. The diameter of the capillary column should be smaller than the diameter of the particle of the sorbent in the microcolumn [26]. Considerable attention was therefore paid to the preparation of columns with a particle diameter of less than 30 μm [54–57] as well as a stationary phase fixed to the wall of the capillary column [58]. The majority of procedures were taken from gas chromatography [59]. However, the expected results have not always been reached. Liquid-solid [52] and liquid-liquid systems were studied, and attention was drawn to the basic fact [60] that, as a result of Laplace-Marangoni flow of the stationary phase, liquid-liquid systems are not suitable for capillary chromatography. Simultaneously we pointed out the restricted use of gas-liquid systems in capillary chromatography [60,61].

The attempt to speed the analysis and to increase column efficiency leads to a reduction in the diameter of the grain of the sorbent or to a reduction in the cross section of capillary columns. This is accompanied by decreasing volumes of the analyte-containing effluent. Sorbents with a particle size of 3 μm and smaller [62,63] are used in analytic columns [64–69] and even in microcolumns [43,45]. The cross section of capillary columns ranges from 5 [52] to 30 μm [58]. This development led to higher standards for minimization of extracolumnar spaces [70]. At present it is not the dimensions of columns that are constraints as far as the efficiency and speed of analysis are concerned, but the minimization of extracolumnar spaces [8,26,71], particularly detector volumes, and suitable solution of the injector.

The size of the detection cell is reduced and the basic detection parameters, such as the response speed, the minimum detectable concentration, and the minimum detectable mass flow rate, are maintained. Spectrophotometric and fluorometric detectors are usually an acceptable compromise [15,72–76]. However, for both microcolumns and capillary columns, electrochemical detectors [41,76,77] and some types of transport detectors, such as ionization detectors based on flame ionization [54,78,79], have been shown to be more advantageous. Detectors with the smallest cell volumes could be designed successfully on electrochemical principles. Potentiometric detectors with cell volumes of the order of femtoliters [80], amperometric detectors with a cell volume of less than 1 nl [81], and conductometric detectors [77,82] with a cell volume below 0.1 μl have been described. Electrokinetic detectors have also been coupled to microcolumns [38].

An independent chapter in the present development of detectors is represented by prospective combinations of microcolumns with spectrometric methods, such as micropacked column mass spectrometry [83] and micropacked column infrared spectrometry [84].

Trace analysis is a significant field for applications of liquid chromatography. Owing to its high mass sensitivity, the technique of microcolumn chromatography became an important tool of modern analytic chemistry. Trace analysis can be effected successfully only by a balanced combination of three procedures: (1) sample enrichment with analyte, (2) low-dispersion separation of components in the chromatographic column, and (3) selective detection [85]. For potential automation of the process of trace analysis it is advantageous, in some instances even necessary, to couple the enrichment process to separation in a chromatographic column [86,87]. The advantages of this procedure [51,76,88–90] were demonstrated for microcolumns.

Sample enrichment with solute is based on the application of a suitable gradient of the elution strength of the mobile phase [91]. A procedure for a preparation using the injection-generated mobile-phase gradient [51,89,90] was proposed. This technique simplifies the instrumentation necessary for obtaining a reproducible gradient and, at the same time, allows analysts to reach very low minimum detectable concentrations, of the order of ng/L. The injection-generated gradient further introduces a new element into chromatography that makes it possible to control separation selectivity by modifying the sample under analysis.

Microcolumn liquid chromatography has definitely proceeded beyond the research and development stage and now ranks among analytic chromatographic methods. In contrast, capillary liquid chromatography has remained in research and development, and unsolved technical problems prevent capillary columns from being widely used in liquid chromatography.

REFERENCES

1. Hamilton P. B., Bogue D. C., Anderson D. C.: Anal. Chem. 32, 1782 (1960).
2. Giddings J. D.: Anal. Chem. 35, 2215 (1963).
3. Huber J. F. K.: J. Chromatogr. Sci. 7, 85 (1969).

4. Huber J. F. K., Hulsman J. A. R.: Anal. Chim. Acta 38, 305 (1967).
5. Kirkland J. J.: J. Chromatogr. Sci. 7, 7 (1969).
6. Golay M. J. E., in: Gas Chromatography 1958 (D. H. Desty, ed.), Butterworths, London, 1958.
7. Martin A. J. P.; in: Gas Chromatography 1962 (M. van Swaay, ed.), Butterworths, London, 1962.
8. Knox J. H.: J. Chromatogr. Sci. 18, 453 (1980).
9. Sagliano N. Jr., Hsu S.-H., Floyd T. R., Raglione T. V., Hartwick R. A.: J. Chromatogr. Sci. 23, 238 (1985).
10. Horvath C., Preis B. A., Lipsky S. R.: Anal. Chem. 39, 1422 (1967).
11. Horvath C., Lipsky S. R.: Anal. Chem. 41, 227 (1969).
12. Nota G., Marino G., Buonocore V., Ballio A.: J. Chromatogr. 46, 103 (1970).
13. Dewaele C., Verzele M.: J. High Resol. Chromatogr. Chromatogr. Commun. 1, 174 (1978).
14. Hibi K., Ishii D., Fujishima I., Takeuchi T., Nakanishi T.: J. High Resol. Chromatogr. Chromatogr. Commun. 1, 21 (1978).
15. Tsuda T., Hibi K., Nakanishi T., Takeuchi T., Ishii D.: J. Chromatogr. 158, 227 (1978).
16. Tsuda T., Novotny M.: Anal Chem. 50, 632 (1978).
17. Tsuda T., Novotny M.: Anal. Chem. 50, 271 (1978).
18. Krejčí M., Šlais K., Tesařík K.: J. Chromatogr. 149, 645 (1978).
19. Ishii D., Hibi K., Asai K., Jonokuchi T.: J. Chromatogr. 151, 147 (1978).
20. Ishii D., Hibi K., Asai K., Nagaya M.: J. Chromatogr. 152, 341 (1978).
21. Ishii D., Hibi K., Asai K., Nagaza M., Mochizuki K., Mochida Y.: J. Chromatogr. 156, 173 (1978).
22. Ishii D., Hirose A., Hibi K., Iwasaki Y.: J. Chromatogr. 157, 43 (1978).
23. Tijssen R.: Separ. Sci. Technol. 13, 681 (1978).
24. Šlais K., Krejčí M.: J. Chromatogr. 148, 99 (1978).
25. Krejčí M., Tesařík K.: AO 167,052; U.S. Patent 4,014,793.
26. Knox J. H., Gilbert M. T.: J. Chromatogr. 186, 405 (1979).
27. Scott R. P. W., Kucera P.: J. Chromatogr. 125, 251 (1976).
28. Scott R. P. W., Kucera P.: J. Chromatogr. 169, 51 (1979).
29. Ishii D., Asai K., Hibi K., Jonokuchi T., Nagaya M.: J. Chromatogr. 144, 157 (1977).
30. Takeuchi T., Ishii D.: J. Chromatogr. 190, 150 (1980).
31. Takeuchi T., Ishii D.: J. Chromatogr. 213, 25 (1981).
32. Takeuchi T., Ishii D.: J. Chromatogr. 213, 469 (1981).

33. Takeuchi T., Ishii D.: J. Chromatogr. 238, 409 (1982).
34. Takeuchi T., Ishii D.: J. Chromatogr. 239, 633 (1982).
35. Hirata Y., Jinno K., in: Microcolumn Separations J. Chromatogr. Library, Vol. 30 (M. Novotny and D. Ishii, eds.), Elsevier, Amsterdam, (1985), p. 45.
36. Yang F. J.: J. Chromatogr. 236, 265 (1982).
37. Gluckman J. C., Hirose A, McGuffin V., Novotny M.: Chromatographia 17, 303 (1983).
38. Krejčí M., Kouřilová D., Vespalec R.: J. Chromatogr. 219, 61 (1981).
39. Krejčí M., in: Interan 1982 (J. Zýka, ed.), CSVTS, Usti nad Labem, Czechoslovakian 1982, p. 18.
40. Kouřilová K., Šlais K., Krejčí M.: Collect. Czech. Chem. Commun. 49, 764 (1984).
41. Krejčí M., Šlais K., Kouřilová D.: Chem. listy 78, 469 (1984).
42. Vespalcová M., Šlais K., Kouřilová D., Krejčí M.: Česk. Farm. 33, 287 (1984).
43. Kouřilová D., Šlais K., Krejčí M.: Chromatographia 19, 297 (1984).
44. Takeuchi T, Ishii D, Nakanishi A.: J. Chromatogr. 285, 97 (1984).
45. Jinno K.: J. High Resol. Chromatogr. Chromatogr. Commun. 7, 66 (1984).
46. Krejčí M., Kahle V.: Czechoslovak Patent 257657.
47. Krejčí M., Kahle V.: J. Chromatogr. 392, 133 (1987).
48. Novotny M., in: Microcolumn Separations, J. Chromatogr. Library, Vol. 30 (M. Novotny and D. Ishii, eds.), Elsevier, Amsterdam, 1985, p. 19.
49. Hirose A., Wiesler D., Novotny M.: Chromatographia 18, 239 (1984).
50. Erni F.: J. Chromatogr. 282, 371 (1983).
51. Krejčí M., Šlais L., Kunath A.: Chromatographia 22, 311 (1986).
52. Krejčí M, Tesařík K, Pajurek J.: J. Chromatogr. 191, 17 (1980).
53. Tijssen R., Bleumer J. P. A., Smith A. L. C., van Krevel O. M. E.: J. Chromatogr. 218, 137 (1981).
54. Krejčí M., Tesařík K., Rusek M., Pajurek J.: J. Chromatogr. 218, 167 (1981).
55. Tesařík K.: J. Chromatogr. 191, 25 (1980).
56. Ishii D., Takeuchi T.: J. Chromatogr. Sci. 18, 462 (1980).
57. Jorgenson J. W., Guthrie E. J.: J. Chromatogr. 255, 335 (1983).
58. Ishii D., Takeuchi T.: J. Chromatogr. Sci. 22, 400 (1984).
59. Tesařík K., Komárek K.: Kapilarní kolony v plynové chromatografii, SNTL, Praha, 1984.

60. Krejčí M., Tesařík K.: J. Chromatogr. 282, 351 (1983).
61. Krejčí M., Tesařík K., Brezina V.: Liquid Stationary Phase Distribution in Capillary Liquid Chromatography, Workshop on Microcolumn Liquid Chromatography, Free University of Amsterdam, 1984.
62. Unger K. K., Messer W., Krebs K. I.: J. Chromatogr. 149, 1 (1978).
63. Dewaele C., Verzele M.: J. Chromatogr. 282, 341 (1983).
64. DiCesare J. L., Dong M. W., Atwood J. G.: J. Chromatogr. 217, 369 (1981).
65. Cooke N. H. C., Archer B. G., Olsen K., Berick A.: Anal. Chem. 54, 2277 (1982).
66. Verzele J., van Dijck J., Mucshe P., Dewaele C.: J. Liquid Chromatogr. 5, 1431 (1982).
67. Unger K. K., Anspach B.: Trends Anal. Chem. 6, 121 (1987).
68. Katz E., Scott R. P. W.: J. Chromatogr. 253, 159 (1982).
69. Kahle V., Krejčí M.: J. Chromatogr. 321, 69 (1985).
70. Šlais K., Kouřilová D.: J. Chromatogr. 258, 57 (1983).
71. Hupe K. P., Jonker R. J., Rozing G.: J. Chromatogr. 285, 253 (1984).
72. Ishii D., Tsuda T., Hibi K., Takeuchi T., Nakanishi T.: J. High Resol. Chromatogr. Chromatogr. Commun. 2, 341 (1979).
73. Hershberger L. W., Callis J. B., Christian G.D.: Anal Chem 51, 1444 (1979).
74. Hirata Y., Novothy M.: J. Chromatogr. 186, 521 (1979).
75. Yang F. J.: J. High Resol. Chromatogr. Chromatogr. Commun. 4, 83 (1981).
76. Krejčí M., Šlais K., Kouřilová D., Vespalcová M.: J. Pharm. Biomed. Anal. 2, 197 (1984).
77. Šlais K.: J. Chromatogr. Sci. 24, 321 (1986).
78. McGuffin V. L., Novotny M.: Anal. Chem. 53, 946 (1981).
79. Krejčí M., Rusek M., Houdková J.: Collect. Czech. Chem. Commun. 48, 2343 (1983).
80. Manz A., Froebe Z., Simon W., in: Microcolumn Separations, J. Chromatogr. Library, Vol. 30 (M. V. Novotny and D. Ishii, eds.), Elsevier, Amsterdam, 1985.
81. Šlais K., Krejčí M.: J. Chromatogr. 235, 21 (1982).
82. Kouřilová D., Slais K., Krejčí M.: Collect. Czech. Chem. Commun. 48, 1129 (1983).
83. Lee E. D., Henion J. D.: J. Chromatogr. Sci. 23, 253 (1985).
84. Taylor L. T.: J. Chromatogr. Sci. 23, 265 (1985).
85. Frei R. W., Brinkman U. A. T.: Trends Anal. Chem. 1, 45 (1981).

86. Jandera P., Churáček J.: Gradient Elution in Column Liquid Chromatography, J. Chromatogr. Library, Vol. 31, Elsevier, Amsterdam, 1985.
87. Huber J. F. K., Becker R. R.: J. Chromatogr. 142, 765 (1972).
88. Šlais K., Kouřilová D., Krejčí M.: J. Chromatogr. 282, 363 (1982).
89. Šlais K., Krejčí M., Chmeliková J., Kouřilová D.: J. Chromatogr. 388, 179 (1987).
90. Šlais K., Krejčí M., Kouřilová D.: J. Chromatogr. 352, 179 (1985).
91. Gluckman J. C., Shelley D. C., Novotny M. V.: Anal. Chem. 57, 1546 (1985).

3
Microcolumns

The problems of microcolumns have been treated in a number of papers [1–8] and monographs [9–14]. Their analytic advantages are accompanied by the known technical problems, which are gradually being solved, however. Solution of the problems connected with minimization of extracolumnar spaces, as well as with the design of columns and detectors, is based on relatively simple theoretical laws of selected solute migration through the chromatographic column.

3.1 THEORETICAL BASES

3.1.1 Influence of Column Diameter on the Concentration Profile of the Solute at the Column Outlet

For estimation of the influence of the chromatographic column inside diameter d_c on the concentration of the solute in the peak maximum at the column outlet c_{max}, let us examine the known relationship that applies, supposing that the peaks at the column outlet have the character of Gaussian curves:

$$c_{max} = \left(\frac{2}{\pi}\right)^{3/2} \frac{Q}{\varepsilon_u (h d_P)^{1/2} \sqrt{L} d_c^2 (1 + k)}$$

$$= \left(\frac{2}{\pi}\right)^{3/2} \frac{Q}{\varepsilon_u \sqrt{L} d_c^2 \sigma_t u} \qquad (3.1)$$

where

h = reduced height equivalent of the theoretical plate

d_P = mean diameter of the sorbent particles

L = column length

k = capacity factor

Q = amount of the solute injected on the column

ε_u = column interparticle porosity

u = mobile-phase linear velocity

σ_t = standard deviation expressed in units of time and characterizing peak spreading on the column

It is evident from Equation (3.1) that the solute concentration in the peak maximum at the column outlet is proportional to the square of the column diameter. This is advantageous from the point of view of column function, and therefore it is an important virtue of micro-columns. It has been proved experimentally [15,16] that the column diameter does not seriously affect the height equivalent of the theoretical plate. This is also proved by the fact that the basic quantity influencing the diffusion path of the solute is the particle diameter d_P. On the contrary, it is evident that by using a sorbent with a smaller particle diameter it is possible to obtain a higher number of theoretical plates while preserving the column length because the value $h \simeq 2$–3 remains unchanged for particles with $d_P = 3$ µm and above.

In Equation (3.1) $h = f(u)$ (also $\sigma_t = f(u)$, which means that low values of h (e.g., $h = 2$) can be obtained only at the optimum velocity of the mobile-phase flow rate in the column. Therefore also $c_{max} = f(u)$. Thus to obtain a lower minimum detectable mass as well as greater separation efficiency of the column or to reduce the time of the analysis it is necessary to optimize the mobile-phase velocity.

These data demonstrate some obvious advantages of microcolumns compared with the conventional columns used at present ($d_c = 4$ mm). In some special analytic tasks microcolumns are even irreplaceable. The advantages of microcolumns are most evident in the trace analysis of extremely small sample volumes (mass). In the chromatographic process, as has been shown, the solute is less diluted with the mobile phase and, consequently, the solute concentration at the column outlet is higher than that obtained in conventional analytic columns.

The quantity Q in Equation (3.1) can be expressed as a function of the solute concentration in the sample c_S and the sample volume V_S injected on the chromatographic column:

$$Q = c_s V_s \tag{3.2}$$

The ratio of the concentration in the solute peak maximum beyond the column c_{max} to the solute concentration in the sample c_s characterizes the degree of analyte dilution during separation on a chromatographic column. In the isocratic mode the relationship $0 < c_{max}/c_s \leq 1$ applies for this ratio. It is evident that the degree of solute dilution also depends on the chromatographic column diameter. The dependence $c_{max}/c_s = f(d_c)$ on the usual values of the quantities in Equation (3.1) is evident from Table 3.1. In an approximate calculation of the degree of analyte dilution during chromatographic separation, the condition $V_s < V_M$ must be applied, and consequently, the quantities L and d_c cannot be chosen at random. However, Table 3.1 shows that by shortening the column length and diminishing its diameter while preserving the values of the other quantities the degree of solute dilution decreases, as does the minimum detectable amount of solute.

3.1.2 Mobile-Phase Velocity

The optimal velocity of the mobile phase u_{opt}, corresponding to the minimum value of the theoretical plate height equivalent, is a function of the molecular diffusion coefficient B and the coefficient C of mass transfer resistance between the phases: $u_{opt} = B/C^{1/2}$. The chromato-

Table 3.1 Ratio of Solute Concentration in Peak Maximum c_{max} to Solute Concentration in the Sample c_s for Columns of Different Diameters d_c and Lengths L^a

Column Diameter d_c (mm)	c_{max}/c_s		
	$L = 100$ mm	$L = 50$ mm	$L = 30$ mm
0.2	7.26×10^{-2}	1.03×10^{-1}	1.32×10^{-1}
0.5	1.16×10^{-2}	1.64×10^{-2}	2.12×10^{-2}
1.0	2.90×10^{-3}	4.10×10^{-3}	5.30×10^{-3}
5.0	1.16×10^{-4}	1.64×10^{-4}	2.10×10^{-4}

^{a}Values of further quantities in Equation (3.1): $h = 2$, $d_P = 5 \times 10^{-6}$ m, $\varepsilon_u = 0.7$, $k = 4$, $V_s = 2 \times 10^{-11}$ m^3 2 $\times$ 10^{-2} µl [according to Eq. (3.1)].

graphic column diameter does not appear in any of the coefficients given, and therefore the mobile-phase optimum velocity is not dependent on it.

The mobile-phase volume flow rate F_m in a packed column may be expressed as a function of the linear velocity by the relationship

$$F_m = \frac{\pi}{4} \varepsilon_u d_c^2 u \qquad (3.3)$$

At the same linear velocity of the mobile phase the volume flow rate is a function of the column diameter. When the column diameter changes from 4 to 0.7 mm (or to 0.5 mm), the mobile-phase volume flow rate at constant linear velocity decreases 33 times (or 64 times).

The pressure drop in the column Δ_P can be described by the equation

$$\Delta_P = \frac{u\eta L}{d_P^2} \frac{\varepsilon_u}{\varepsilon_u + \varepsilon_i} = \frac{4}{\pi} \frac{F_m \eta L \varphi}{d_c^2 d_P^2 (\varepsilon_u + \varepsilon_i)} \qquad (3.4)$$

where

η = mobile-phase dynamic viscosity

ε_i = intraparticle porosity

φ = coefficient of the column hydrodynamic resistance

$$\varphi = \frac{180\psi^2(1 - \varepsilon_u)^2(\varepsilon_u + \varepsilon_i)}{\varepsilon_u^3} \tag{3.5}$$

ψ^2 is a coefficient characterizing the shape of the sorbent particles.

Equation (3.4) shows that the pressure drop in the column is independent of the column diameter. If the same column length, packing, grain size, mobile phase, and linear velocity are used, a column of small diameter will have the same pressure drop as a conventional column. Such a theoretical conclusion, however, applies only if all the quantities in the equation are really not dependent on the chromatographic column diameter. For ε_u, however, this need not be true. It has been found experimentally [17], for example, that to obtain the same linear velocity under conditions that are identical in all other respects, a lower pressure is necessary for microcolumns compared to columns of conventional diameter. Comparing column diameter $d_c = 4.6$ mm with column diameter $d_c = 0.32$ mm, the pressure drop in a conventional column with sorbent diameter $d_P = 10$ μm is 1.77 times higher than that in a microcolumn with the same sorbent. If the columns are packed with sorbent of diameter $d_P = 1$ μm, the pressure ratio in both columns equals 1.97. These results prove that microcolumns have a higher permeability than columns of conventional diameter. If the limiting factor is the maximum pressure obtained, longer (and hence more effective) microcolumns can be used instead of conventional columns. This conclusion, however, applies not only for analytic application of the columns but also for their preparation. Microcolumns can be packed with sorbents of 3 μm particle diameter (or less) using the same pressure necessary for packing conventional columns while preserving sufficient velocity of packing, which is proved by the fact that they reach theoretical efficiency.

3.1.3 Length of the Microcolumn

The number of theoretical plates necessary to obtain the required separation of the sample components is the basic criterion determining the length of the chromatographic column. In present-day microcolumn liquid chromatography, two main types of columns are used: (1) long columns designed to obtain a large number of theoretical plates and (2) short columns enabling the rapid analysis of less complex mixtures. Long columns are characterized by increasing pressure drop, short columns by increasing demands on minimization of extracolumn spaces.

Scott and Kucera [18] connected several stainless steel microcolumns of internal diameter 1 mm and 1 m long, packed with silica gel of mean particle diameter 5 μm. They obtained 650,000 theoretical plates on such a column that was 14 m long. Packed microcolumns several meters long were also prepared with thin-walled glass capillaries of 0.08–0.25 mm diameter and capillaries from fused silica [19] of diameter 0.057–0.376 mm. These columns also had high efficiencies (from 100,000 to 200,000 theoretical plates).

For most chromatographic separations from 5,000 to 10,000 theoretical plates is sufficient [20]. With respect to the large variation in selectivity, a considerable number of separations can be solved even on columns with a number of theoretical plates not exceeding 5,000. If according to Equation (3.6) $L = hd_pn$, the required number of theoretical plates can be obtained with relatively short columns. For $h = 2$ the following values are obtained:

d_p, μm	3	5	10
L, mm	30	50	100
n	5,000	5,000	5,000

Therefore, considerable attention has been paid to the study of the properties of short columns [16,21–24], and the possibility of different types of analysis has been demonstrated [24,25].

3.1.4 Retention Volume

The most important absolute retention characteristic in chromatography, retention volume V_R, also depends on column diameter:

$$V_R = V_M(1 + k) = \frac{\pi}{4} d_c^2 L\varepsilon_u(1 + k) \tag{3.6}$$

The decrease in the retention volume due to lower values of d_c enables us to solve in a technically new way the mobile-phase pumping problem [26] (see Sec. 3.2) and also to find a mobile phase matched to a given analyzed sample; that is, we can now include the chemical properties of the mobile phase in the chromatographic analysis [24].

Preparation of a suitable mobile phase as well as derivatization of the solute either before its separation on the column [7,27] or after separation (before it enters the detector) significantly affects the migration of the solute on the column. The composition of the mobile phase and the contribution of modifiers of its elution strength can adjust the value of the retention volume or of the solute capacity ratio (see Sec. 4).

It can also be proved that even relatively low absolute values of retention volume are sufficient both for routine analysis and for chromatographic separation of the components of more complex mixtures.

For the dependence of basic chromatographic parameters on the column diameter at constant V_R, the following equations can be derived [24]:

$$n = V_R[0.25\pi d_c^2 \, h d_p \varepsilon_u (1 + k)]^{-1} \tag{3.7}$$

$$k = V_R[0.25\pi d_c^2 \, L\varepsilon_u]^{-1} - 1 \tag{3.8}$$

$$L = V_R[0.25\pi d_c^2 \, \varepsilon_u (1 + k)]^{-1} \tag{3.9}$$

Table 3.2 presents an example of parameters obtainable at $V_R = 100$ μl.

The required number of theoretical plates n_{req} can be derived similarly as a function of the column diameter for the given resolution, R_s, from the relative retention (retention ratio) $r_{1,2}$ in the form

$$n_{req} = 16R_s^2 \left(\frac{r_{1,2}}{r_{1,2} - 1}\right)^2 \frac{\omega^2}{d_c^4} \left(\frac{\omega}{d_c^2} - 1\right)^{-2} \tag{3.10}$$

where $r_{1,2} = k_2/k_1$ and $\omega = 4/\pi(V_R/\varepsilon_u L)$. It is evident from the form of Equation (3.10) that $\omega/d_c^2 \neq 1$ and simultaneously $r_{1,2} \neq 1$, which means that $k > 0$ and therefore

$$L > L_{lim} = \frac{4V_R}{\varepsilon_u d_c^2 \, \pi} \tag{3.11}$$

Table 3.2 Values of Some Chromatographic Parameters
Obtainable in One Stroke of the Injection Piston (Injection Cylinder
Volume 100 μl)

	Chromatographic Column Inside Diameter (cm)			
	0.02	0.05	0.1	0.2
Number of theoretical plates,[a] n	9.1×10^4	1.5×10^4	3.6×10^3	9.1×10^2
Column length L,[a] cm	90.9	14.5	3.6	0.9
Capacity ratio,[b] k	30	4.7	1.1	0.2

[a] $k = 4$, $h = 2$, $d_P = 5$ μm.
[b] $L = 15$ cm.

The limiting length of the column L_{lim} represents the value at which $V_R = V_M$.

The possibilities for microcolumn chromatography with the low retention volume $V_R = 100$ μl are shown in Figure 3.1. It is supposed here that

$$H_r = \frac{L}{n_{\text{req}}} = h d_{P,\text{req}} \tag{3.12}$$

where H_r is the height equivalent of the theoretical plate necessary for the required separation and $d_{P,\text{req}}$ is the sorbent particle diameter necessary for the given separation.

Under the condition $h = 2$, the necessary height equivalent of the theoretical plate H_r according to Equation (3.12) corresponds to double the value of the sorbent particle diameter $d_{P,\text{req}}$. The lower the value of d_P that is selected, the shorter the column needed for the required separation. The limiting diameter of the sorbent particle at present is $d_P \simeq 2\text{-}3$ μm. This consideration, illustrated by Figure 3.1, does not take into account the pressure drop in the column. The necessary

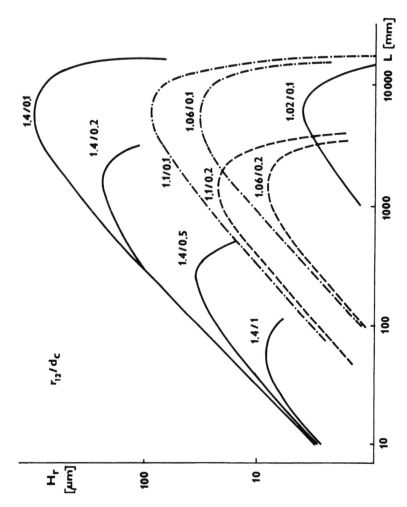

Figure 3.1 Dependence of $H_{req} = L/n_{req}$ on L for the given relative retention $r_{1,2}$ using a column of given inside diameter d_c (values $r_{1,2}/d_c$ are expressed by the numbers on the curves); $V_R = 100$ μl.

pressure drop in the column increases with decreasing particle diameter of the sorbent at a constant column length and mobile-phase velocity.

The data given show that only a relatively very small retention volume in microcolumn liquid chromatography is sufficient for considerably demanding separation of pairs of solutes (see, e.g., Fig. 3.15). Perhaps in the future this can be used for the design of considerably simplified and at the same time highly automated and universally applicable chromatographic devices.

3.1.5 Injected Sample Volume

With respect to the distortion of the chromatographic zone, the maximum applicable volume of the injected sample depends on the number of column theoretical plates n and on the retention volume V_R:

$$V_s = \frac{aV_R}{n^{1/2}} \tag{3.13}$$

where a is a constant that depends on the permissible extent of peak spreading. With regard to the additivity of individual contributions to spreading of the chromatographic curve, the volume variations must be

$$\sigma_s^2 + \sigma_{v^\circ C}^2 = a\sigma_{v^\circ C}^2 \tag{3.14}$$

where σ_s^2 is the variance in solute volume spreading caused by the injected sample volume and $\sigma_{v^\circ C}^2$ is the variance in nonsorbed solute volume spreading in the column.

If the distortion is not to exceed 10%, then $a = 1.1$ and $\sigma_s^2 \leqslant 0.1\sigma_{v^\circ C}^2$. The variation in spreading the rectangular concentration pulse carrying the sample on the column is given by the relationship

$$\sigma_s^2 = \frac{V_s^2}{12} \tag{3.15}$$

A decrease in the column diameter is accompanied by a decrease in the sample volume that can be injected in the column without losing its efficiency. If the injected sample volume V_s is required to cause maximally a 10% contribution to eluted zone spreading, then

$$\sigma_s^2 = \frac{V_s^2}{12} \leqslant 0.1\sigma_{v^\circ c}^2 \qquad (3.16)$$

When using a column of inner diameter 0.7 mm and 15 cm long packed with sorbent of particle diameter 6 μm, the injected volume should reach a maximum of 0.45 μl; using a column of inner diameter 0.5 mm and 10 cm long the volume can be 0.22 μl. By shortening such a column to only 3 cm, $V_s = 0.1$ μl is obtained. If short quartz capillaries of inner diameter 0.2 mm were used, Equation (3.13) could be fulfilled only under technically impossible conditions.

   There is a disproportion between the small retention volume and the injected sample volume. For analytic application of chromatography with small retention volumes, substantially higher sample volumes would have to be injected than those allowed by Equation (3.13). The volume V_s reaches such values that the difference between V_s and V_R substantially decreases. In this case the elution strength of the sample matrix characterized by the solute capacity ratio k_s begins to act. If the solute capacity ratio in the mobile phase is k and the column dead volume V_M, the process of injection can be considered the formation of a stepwise gradient of mobile-phase strength [27] and for the reduced retention volume V_R' the following relationship applies:

$$V_R' = V_s \frac{k_s - k}{k_s} + kV_M \qquad (3.17)$$

If $k_s = k$, the maximal value of V_s is determined according to Equation (3.13). If $k_s \gg k$, Equation (3.17) changes to

$$V_R' = V_s + kV_M \qquad (3.18)$$

Under such circumstances the injected sample volume can exceed the column dead volume several times.

   From a general point of view solute retention in liquid chromatography is always affected by the volume of the injected sample, whose composition always differs from the composition of the mobile phase. Consequently, in the initial sections of the column, the solute is eluted by a liquid with a composition different from the mobile-phase composition. The influence of the injected sample volume on V_R decreases with decreasing V_s and at the same time with increasing V_M. At constant values of L, k_s, and k, the influence of the injected volume decreases with increasing column diameter, that is, with increasing V_M. Equations (3.17) and (3.18) show that the amount of the injected sample affects not only the column efficiency but the retention characteristics of the solute as well.

3.1.6 Influence of Sorbent Particle Size

The sorbent particle diameter d_P is the quantity determining the length of the diffusion path of the solute in the chromatographic column. Therefore, it decisively influences chromatographic peak spreading. Decreasing peak spreading is connected with decreasing particle diameter of the chromatographic packing. The values of d_P currently used decreased from the original at about 50 to 10 μm or most often 5 μm. Ever-increasing demands on the speed of analysis led to preparation and applications of sorbents with a grain size of 3 μm and less [22,28–34].

Selection of small d_P enables us not only to obtain a higher number of theoretical plates at a constant column length but also to carry out analyses within a few seconds [29]. If a sorbent with small particles is used, the mobile-phase optimal velocity (e.g., Ref. 35) is in the region of higher values, and moreover, the minimum of the dependence $H = f(u)$ is very flat [28,29]. However, a disadvantage of a sorbent with small values of d_P is a higher pressure drop in the chromatographic column, especially a very small mobile-phase volume in which the solute is eluted from the column. For the nonsorbed component the volume peak variation $\sigma^2_{V^\circ C}$ can be derived from the basic equation for the number of theoretical plates on the column:

$$\sigma^2_{V^\circ C} = \frac{V_M^2}{n} = \frac{\pi^2}{16} d_c^4 \varepsilon_u^2 L h d_P \qquad (3.19)$$

It is evident that the volume variation is a function not only of sorbent particle diameter but also of chromatographic column diameter. The demands for the minimization of extracolumnar spaces considerably increase. Therefore, for short microcolumns and less sorbed solutes, grains of greater diameter ($d_P = 5$–10 μm) are recommended.

3.2   APPARATUS

Microcolumn liquid chromatographs consist of three basic components:

1.   Supply of constant flow of the mobile phase, in some cases supplemented with a device for forming a gradient of mobile-phase composition
2.   Columns, including the injection device
3.   Detector and signal processing

All the elements of the chromatograph must be adjusted to the requirements of miniaturization, which are more difficult to fulfill for some parts of the device and less for others. This is also reflected in the range and arrangement of this section. Greater attention is paid to parts that are miniaturized with difficulty. Because sample injection is considered an important process directly connected with trace analysis, the section dealing with sample injection follows the section on detectors.

3.2.1   Pumps

The value of the retention volume as the basic information on solute quality is found by determining the retention time of the solute at a constant mobile-phase flow rate. Mobile-phase flow through the column in microcolumn chromatography is enabled by pumps whose design and working principles do not much differ from those of pumps suitable for conventional liquid chromatography.

We usually meet with pumps that provide constant flow of the mobile phase as well as those that provide constant pressure of the mobile phase. Practically speaking, in terms of retention it is more advantageous to use pumps that provide constant flow of the mobile phase regardless of column resistance. Column resistance can be altered by chance, for example, by changing the position of the sorbent in the column or by applying impurities to the column inlet, or systematically, for example by a change in mobile-phase viscosity as a result of a change in its composition. It is important to stress, however, that the design of pumps that provide the reduced constant flow required in microcolumn chromatography is extremely demanding.

These pumps reach a maximum working pressure from 30 to 40 MPa. The mobile-phase flow rate is in the range from 1 to 500 μl/min. Many pumps are designed from chemically inert materials: glass, Teflon, titanium, and the like. The chemical inertness and, consequently, reduced catalytic activity of the construction materials is an important criterion in microcolumn chromatography; microcolumn liquid chromatography is well suited to use in biochemical and medical analyses, in which chemically unstable substances are often investigated. A further important parameter is the period during which the pump achieves the working pressure. This parameter is especially important for the "stop-flow" technique of injection, that is, sampling with interrupted flow of the mobile phase. It is clear that without suitable pump regulation rapid adjustment of the pressure would be very difficult.

Microcolumn chromatographs as well as conventional liquid chromatographs are used with three basic types of pumps: (1) pneumatic pumps, (2) reciprocal pumps, and (3) syringe pumps. Pneumatic pumps often use pneumatic amplifiers, and they are of the simplest design. However, they work only as the flow supplied in the regime of constant pressure. They are used in microcolumn chromatography only rarely.

Most frequently reciprocal pumps are used. Modern regulation of the flow rate and application of suitable stepping motors enable us to reduce the mobile-phase flow rate to 1 μl/min. Reciprocal pumps provided with a working cylinder of around 0.1 ml volume work in many cases like syringe pumps because the volume of the liquid pushed out on one stroke of the piston substantially exceeds the retention volume of the analyte many times. Problems often arise as a result of the tightness of check valves, mostly those of the ball type. These need not work reliably at very low flow rates because the pump efficiency changes and the liquid flow is not constant. A solution is the forced control of the pump check valve. The minimum volume obtainable at one step of the motor is also important. At badly selected pump cylinder dimensions the volume often reaches 10 μl or more. Under such circumstances any further decrease in the stepping motor frequency results in liquid flow pulses that are mostly harmful to the function of the detector. This defect is solved by inserting into the pump a flow splitter, which reduces the flow rate of the liquid that is pushed out.

For columns of diameter below 0.7 mm syringe pumps seem to be the most suitable. Their cylinders are usually of 5–10 ml volume and they are made of glass, metal coated with Teflon, titanium, or stainless steel. The volume of the ejected liquid depends on the velocity of the piston stroke. The pistons are usually made of glass or steel covered with plastic. The piston stroke is produced by a stepping motor with adjustable velocity. The total volume of the cylinder usually many times exceeds the retention volume of the strongest sorbed analyte. Therefore, the velocity of the cylinder repacking is not important. Packing usually takes several tens of seconds. The inlet and the outlet of the liquid from the pump must be provided with stop valves. When both the valves are closed, the necessary pressure is obtained in the pump cylinder. After opening the valve the pressure is transferred to the column. Despite its obvious advantages syringe pumps also have some shortcomings. Liquid compressibility may influence the precise measurement of retention characteristics. Sometimes problems are also caused by their design. Especially with pumps with a non-metal cylinder and piston, it is difficult to empty the cylinder completely because the piston in the upper dead center does not touch the face of the cylinder. Consequently, it is necessary to test the composition of each new mobile phase for contamination by the phase used before. Despite these problems, syringe pumps can be considered standard for microcolumn liquid chromatography.

3.2.2 Formation of the Mobile-Phase Gradient

Gradient techniques are currently used in liquid chromatography for regulating the retention of the solutes of complex mixtures. Devices for the formation of the mobile-phase gradient in microcolumn chromatography can be divided into three basic categories:

1. The device working on the low-pressure part of a pump
2. The device working on the high-pressure parts of two pumps
3. The device working on the high-pressure part of a single pump

The first type, the low-pressure gradient device, works only in connection with reciprocal pumps. On the inlet part of the pump two or more closing inlet valves are placed, each of them being opened for a certain time by an electronic control device. The liquids are already partly mixed in the working cylinder of the pump at a suitable ratio between the frequency of opening the inlet valves and the piston stroke. However, chromatographs with this type of gradient preparation must still be provided with a mixer to ensure proper mixing of the liquids that form the gradient. For low mobile-phase flow rates the mixer together with the capillary supplying the mobile phase to the injector must be of a very small volume not exceeding the dead volume of the column. That the dead volumes of microcolumns range from less than 10 to several tens of microliters suggests how demanding it is to solve the problem technically. A number of suitable T shapes were designed to ensure proper mixing of the liquids. However, a reliable device has not yet been designed. The essential requirement for proper function of the device is also faultless function of the inlet valves. They must be perfectly tight with a constant pressure drop. The virtue of this type of the device is the possibility of preparing ternary and multicomponent mixtures.

The second type of device works in the high-pressure area of the microcolumn chromatograph. It consists of two pumps of the reciprocal or syringe type connected into a regulation loop that ensures a programmable flow rate of both components of the mobile phase. The design of this type of gradient device is also demanding because the total constant flow of the mobile phase must be kept very low. The problem of mixing, which is mostly performed in conventional chromatographs by combining static and dynamic mixers, also becomes in high-pressure gradient devices considerably difficult technically. In some cases the flow splitter is positioned to follow the mixers so that the liquids can be mixed at higher flow rates and, consequently, under more easily controllable conditions. However, this position of the divider is connected to a loss of one of the basic advantages of microcolumn chromatography, that is, the economy of the mobile phase. The low-pressure gradient device is often proposed as the solution. Modern microprocessor programming enables us to prepare even the complex course of ratio changes in concentrations for a two-component mobile phase. If the mixers function well, the mobile-phase gradient is highly reproducible.

The third type of gradient device works on the high-pressure part of a single pump. It makes use of volume pulse injection of the mobile phase with a higher elution strength than the flow of a weaker mobile phase. The gradient is formed by a six-port injection valve in a loop with the spreading element [36–38]. The pump pushes into the chromatograph a constant flow of the mobile phase with a higher elution strength; the spreading element is packed with the mobile phase with a lower elution strength. By inserting the loop into the mobile-phase flow the mobile phase with the increased elution strength is carried to the injection device and on to the column. The concentration profile at the contact between the two phases is spread, and a monotonous gradient of mobile-phase strength, whose slope can be regulated, is formed. An example of such a device is shown in Figure 3.2a. The concentration profile depends not only on the inside diameter of the capillaries (1 and 2) used for gradient preparation but also on the radius of the coils of the capillaries. First the loop with coils 1 and 2 is packed in the direction from *a* to *b* with the mobile phase of higher elution strength. A certain volume of the mobile phase of lower elution strength is carried into the coils in the direction from *b* to *a*. The weaker mobile-phase volume, which fills the loop, corresponds with the slope of the mobile-phase gradient (see Fig. 3.2b). The limiting values are the elution strength of the mobile phase in the loop and the mobile phase in the pump. Such a mobile-phase gradient has a number of advantages for microcolumns. In any case, the constant flow rate of the mobile phase is maintained. This device requires no independent mixers, whose efficiency is usually considerably dependent on the flow rate value. Despite the inexpensive and simple design of these devices, they can

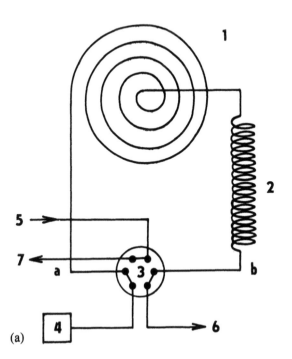

(a)

Figure 3.2 (a) The gradient device according to Reference 37: (1) coiled capillary (inside diameter 1 mm, length 0.8 m); (2) capillary (inside diameter 0.25 mm, length 2 m); (3) six-way valve; (4) pump with final (stronger) mobile phase; (5) initial mobile phase; (6) column; (7) discharge. (b) Examples of chromatograms and mobile-phase gradients prepared in the device according to a. Microcolumn 100 × 1 mm, packed with Spherisorb ODS-2, $d_p = 5$ μm; detector KRATOS 757 UV, 280 nm; sample: solution of chlorophenols in water, volume 5 μl: (1) phenol (62 mg/L); (2) 4-chlorophenol (62 mg/L); (3) 2,4-dichlorophenol (64 mg/L); (4) 2,4,6-trichlorophenol (54 mg/L); (5) 2,3,4,5-tetrachlorophenol (184 mg/L); (6) pentachlorophenol (370 mg/L). Initial mobile phase 30% (vol/vol) of methanol in water; final mobile phase 80% (vol/vol) of methanol in water. Mobile-phase flow rate 1 μl/s. Numbers in chromatograms I–V correspond to the volume (ml) of the initial mobile phase injected in the device according to a. G, start of gradient; I, sample injection.

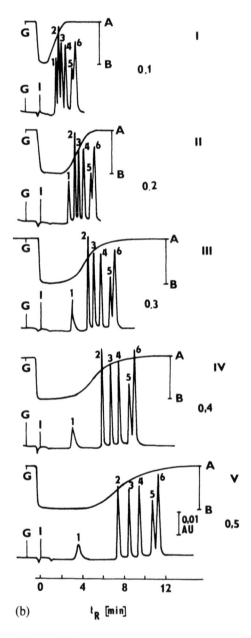

(b) t_R [min]

Figure 3.2 Continued

be successfully automated and used for series analyses. Moreover, they do not contradict the advantage of microcolumn chromatography—reduced consumption of the mobile phases—and they are the most prospective devices for gradient preparation.

3.2.3 Columns

The columns are considered the heart of the chromatograph. Their properties substantially affect the separation of analyzed components, and their diameter affects in a decisive way the detection limit of a chromatographic method.

It has been shown that the diffusion path of the solute in the chromatographic column is a function of the particle diameter of the sorbent. The geometric properties of the sorbent (grain size and shape), together with the arrangement of the column packing, affect the efficiency of the column. The important factors can be divided into two categories. The category that can be directly influenced by the analyst includes the following basic factors:

1. The microcolumn must be packed with a suitable sorbent so that no empty spaces are formed and individual particles cannot change their positions; this prevents a change in chromatographic column efficiency as a result of the destruction of the packing structure by pressure when the sample is packed.
2. Both the injector and the detector must be connected to the microcolumn so that the extracolumnar spaces that cause undesirable peak spreading are minimized.
3. Successful analysis logically depends on the selection of a suitable system of the stationary and mobile phases.

The other category includes the properties of the column that cannot in practice be influenced by the analyst:

1. Sorption properties of the sorbent surface used in the column
2. Size, size distribution, and shape of the sorbent particles

Even though the influence of the quality of the microcolumn capillary surface is not as important as it once seemed to be, the selection of the column material is influenced by the demands of design, such as mechanical strength and pressure stability. Packed columns of small diameter are made of capillaries from poly(tetrafluoroethylene) (PTFE, Teflon), stainless steel, glass, or fused silica. Capillaries of glass [39] and fused silica [40], which have smoother inside walls than capillaries of PTFE and stainless steel, seem to be the most suitable. Thin-walled microcolumns work in a similar way as the glass capillary columns in liquid and gas chromatography. However, it is also possible to design glass columns from thick-walled capillaries [8,16,21] with substantially higher mechanical stability. The thick-walled glass capillaries for microcolumns are stable even without prior treatment of the walls up to a pressure of approximately 60 MPa [41].

The literature presents relatively little information on packing the columns. In all cases the sorbent is packed in the column in suspension. From the large amount of often controversial information on the technique of packing analytic chromatographic columns in suspension, two generally applicable requirements can be stressed [42]. The velocity of injection of the suspension into the column must be so high that segregation of the sorbent according to particle size or material density is prevented. Therefore, a sufficiently high pressure must be used for packing the column to obtain sufficiently rapid deposition of the particles in the column. Even with high linear velocities of the suspension the column can be packed with a relatively small volume flow rate. Therefore, it is possible to work with pumps of only moderate power from the point of view of volume flow rate, merely to enable the necessary pressure to be achieved. The supply of the stationary-phase suspension can be minimized. Consequently, problems with compressibility of the suspension liquid or with expansion of the suspension supply are eliminated.

A graphic representation of the device for packing microcolumns with sorbent is shown in Figure 3.3. In most cases this device can be easily assembled from parts often used in the analytic laboratory. To ensure success it is first necessary to connect all the elements of the device tightly to prevent leakage. In particular the faultless connection of the reservoir of the stationary-phase suspension with the column must be ensured. Other parts are usually connected permanently, and

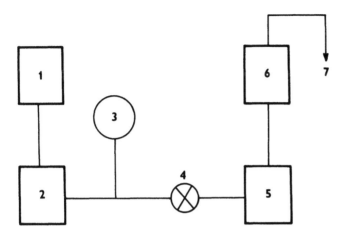

Figure 3.3 The packing device for microcolumns: (1) solvent reservoir; (2) pump; (3) manometer; (4) stop valve; (5) packing suspension reservoir; (6) microcolumn; (7) discharge.

they can be used repeatedly without danger of leaks. When preparing microcolumns it is also important not to forget safety. All the parts of the device and the connections that are under pressure during packing should be placed behind a shield to protect the operator. To ensure rapid injection of the suspension into the column it is necessary to start pump 2 (Fig. 3.3) when valve 4 is closed. Only after the required pressure indicated on manometer 3 is obtained should valve 4 be opened, which will enable sufficiently rapid packing of the microcolumn. After transferring the packing from the stationary-phase reservoir to the column, the liquid in the reservoir is exchanged for a washing liquid. It is useful to wash the column in the end with the mobile phase to be used for the analysis. The completion of stationary-phase transfer from the reservoir to the column is indicated by the manometer. If a pump supplying the device with a constant volume of the liquid is used, the end of the transfer is accompanied by pressure stabilization at the highest value obtained after the pressure drop during the opening of stop valve 4 at the beginning of packing. However, application of this type of pump is connected with undesired changes in the flow rate. Even if these changes are not great because the volume of the system before stop valve 4 (especially manometer 3) is sufficiently large to dampen these variations, they can be the reason for incomplete packing of the column. It is therefore better to work the pumps at a constant pressure. In this case the end of the suspension transfer is indicated by the drop and stabilization of the packing liquid flow rate through the device. When packing is complete, it is necessary to wait for the pressure drop. The column cannot be disconnected earlier because this would cause expansion of the packing and interruption of the continuous sorbent column.

The other general problem necessary to solve in packing columns from the suspension reservoir is particle agglomeration. When preparing a suspension of particles with $d_P < 10$ μm, agglomeration of particles often has a very negative effect. It is supposed [43] that attractive forces originate between the negative centers and positive hydrogen of the acid silanol groups present both on the surface of pure silica gel and on the chemically modified silica gel. The homogeneous charge on the surface of the particles contributes to suppression of these attractive forces. A predominantly positive or negative charge on the silica gel surface is obtained by adjustment of the suspension [16], for example by adding aqueous ammonia, basic alkylamines, strong organic acids, aqueous buffer solution, or nonionogenic tenside. Adding ionogenic tenside also provides good results.

Thick-walled microcolumns can be prepared from a tube of soft glass with inside diameter (ID) 0.5–0.7 mm and outside diameter (OD) 4.5 mm. The glass tubes should be cut to the required length and their ends ground. The appropriate sorbents usually a particle size from 3 to 15 μm. The columns are mostly packed with a suspension of methanol and glycerol (1:1 vol/vol) with 0.1% (wt/vol) sodium dodecyl sulfate. The pressure liquid is methanol. The suitable suspension concentration for particles of 10 μm diameter is 15% (wt/vol); for particles of 5 μm diameter better results can be obtained with a suspension concentration of 10% (wt/vol). The columns are packed at a pressure from 20 to 35 MPa. After packing the column should be washed with methanol and water, preferably in the packing device.

Microcolumns prepared from a thick-walled capillary are often installed in the chromatograph in metal jackets. Metal rings attached to the ends of the columns overlap them by 2–3 mm form a suitable space for a seal. The column is then placed in a metal tube and fixed with two nuts screwed to the ends of the tube. One nut connects the column with the injector; the other connects the column with the detector. The connection can be done simply by capillaries; however, it is preferable to align the inside space of the nut so that it allows direct connection of the injector and the detector with the column. The indisputable advantages of this type of column are ruggedness, easy exchangeability, and simple connection of the column to the chromatograph. Another important feature is the quality of the seal because the tightness of the connection must be maintained even at the relatively low pressure produced by tightening the nuts at the ends of the column jacket. If greater strength must be exerted to achieve a tight seal, the pressure is transferred to the glass of the column, which is already under pressure, and the operating pressure of the liquid may cause the column to break.

Metal screws inserted directly in the ends of the thick-walled column represent another way of connecting the microcolumn to the chromatograph. Connection of the columns with the injector and the detector is usually reliable. The columns must be sealed by tightening the nuts by suitable wrenches. The ruggedness of such glass microcolumns is substantially less than that of the glass columns in metal jackets.

Microcolumns prepared from fused silica capillaries are often used for their undeniable advantages. The outside surface of the columns is usually covered with a polymer, most often polyimide, layer, which makes the fused silica capillaries considerably more mechanically stable. They have long proved their qualities in gas chromatography. The inside diameter, most often around 0.2 mm, allows one to make good use of the advantages of microcolumns. A great advantage also consists in the possibility of selecting a column long enough to obtain the necessary number of theoretical plates without losing the separation efficiency in the connections between individual segments of shorter columns.

If fused silica capillaries are used as packed microcolumns, the problem appears how to equip the column with a seal at the outlet to prevent sorbent leakage from the column both during packing and during analytic operation of the columns. The simplest solution is represented by another two techniques.

The first method of closing the column [44] consists of providing the end of the column with a capillary from 5 to 15 mm long whose inside diameter corresponds to the outside diameter of the column. This can be made of plastic (e.g., PTFE) but also of silica with suitable dimensions. Into this capillary a thin layer (1–3 mm) of silica wadding is crammed. A silica capillary with the same outside diameter as the column but with a substantially smaller inside diameter (10–50 μm) is inserted through the free end to touch the layer of silica wadding. The length of this outlet capillary is selected according to the detector because it forms a connection between the column and the detector. Simple insertion of the capillary is usually tight enough because at the column outlet there is only the pressure drop caused by the hydraulic resistance of the connecting capillary and the detector at a given flow rate. If needed the connections between the capillaries can also be pasted together.

The other possibility used for sealing the columns is insertion of a glass cone [44] at the outlet of the capillary column. A conical hair is drawn from a glass rod. The thinner end of the cone is inserted into the column to a length of about 5 mm, and the overhanging part of the glass cone is broken away. This technique makes use of the fact that the cross section of neither the silica capillary nor the glass rods has the form of a perfect circle. Consequently, on contact of both the bodies tiny holes are created that reliably prevent escape of the particles even with a diameter of 3 μm, allowing, however, for discharge of the liquid. The pressure drop on this seal is not high, and the volumes in the seal are so small that extracolumn spreading is mostly unmeasurable.

One of important advantages of microcolumns prepared from silica capillaries is easy adjustment of the column shape to the needs of the device. Longer columns can be coiled. It is not necessary to change the distance between the injector and the detector or to use longer connecting capillaries. However, the radius of the coil into which the column is wound affects the secondary flow in the column, and it may cause a loss of column efficiency. The minimum allowed radius of this coil depends on both the column diameter and the sorbent particle diameter. For example, for a column with $d_c = 1$ mm and $d_P = 20$ μm a coil radius of 6 cm represents a loss of efficiency of about 10% compared to a column with a coil radius of 11.5 cm [11]. Further diminishing the coil radius would mean an unacceptable decrease in efficiency. A column wound to a coil of radius 1.5 cm has only about 20% of the original efficiency.

It has been stated that packing microcolumns are applicable in all the fields of liquid chromatography. This means that practically all currently prepared sorbents can be used in these columns. Characterization of the surface chemical properties of the sorbents is presented in Table 4.4.

3.2.4 Extracolumnar Spaces

Packed microcolumns enable us to obtain such efficiency that the experimental values of the reduced plate approach theoretical values. With respect to better heat removal the small diameters allow us to use higher pressures than those that can be applied to analytic columns without loss of their efficiency [46,47]. With a packed microcolumn a higher number of theoretical plates can therefore be obtained because a longer column can be used. Dispersion of the solute on the column is characterized by the volume variance $\sigma^2_{V°C}$. For the nonsorbed solute this variance can be expressed by (e.g. Ref. 8)

$$\sigma^2_{V°C} = \frac{\pi^2}{16} h d_P L \varepsilon_u^2 d_c^4 \tag{3.19}$$

It is evident that the variation in the volume of the column decreases with the fourth power of the column diameter. It is also directly proportional to the particle diameter of the sorbent used for separation. A suitable balance of these two quantities forms the basis of modern low-dispersion chromatography. However, it is clear from relationship (3.19) that reduced dispersion is also facilitated by shortening the column or by decreasing the porosity of the chromatographic packing.

For the sorbed solute Equation (3.19) can be written as

$$\sigma_{v,c}^2 = \frac{\pi^2}{16} hd_P L\varepsilon_u^2 \, d_c^4 \, (1 + k)^2 \qquad (3.20)$$

If it is assumed that we work with a column perfectly packed with sorbent (d_P and ε_u), under the given circumstances the value of the reduced plate will also be constant.

However, in liquid chromatography the perfectly packed column is the condition necessary to obtain high efficiencies. The extracolumnar spaces play an important role in the separation efficiency of the chromatographic system, and from the beginning of the revival in liquid chromatography in the late 1960s they have attracted attention. Chromatographic separation is often deteriorated in the extracolumnar spaces: that is, the solute zone separated on the column is additionally dispersed.

Because the migration effects on the chromatographic column have a stochastic character, the resulting variance is given by the sum of variances (e.g., Ref. 48). In a simplified case when all the extracolumnar contributions to dispersion of the chromatographic zone are expressed as $\sigma_{V,ex}^2$, the resulting dispersion of the solute during separation corresponds to

$$\sigma_V^2 = \sigma_{V,c}^2 + \sigma_{V,ex}^2 \qquad (3.21)$$

We investigated the influence of the extracolumnar spaces under conditions of low-dispersion chromatography at higher velocities of the mobile phase [16,29]. We assumed that a great majority of extracolumnar spaces have the character of capillaries with different diameters d_{cap}. It is known that the volume variance of the solute in a capillary of length L_{cap} can be expressed as

$$\sigma_{V,cap}^2 = \frac{\pi^2}{24} \left(\frac{d_{cap}}{2}\right)^6 \frac{u_{cap}L_{cap}}{D_m} \qquad (3.22)$$

where u_{cap} is the linear velocity of the mobile phase in the capillary and D_m the diffusion coefficient in the mobile phase.

For further analysis it an advantage to use Equation (3.20) in the form

$$\sigma^2_{V,c} = \kappa H(1 + k)^2 \tag{3.23}$$

where

$$\kappa = \frac{\pi^2}{16} L d_c^4 \, \varepsilon_u^2 \tag{3.24}$$

The height equivalent of the theoretical plate H can be expressed as a function of the mobile-phase velocity u, for example by the most often used relationship [49,50]

$$H = A + \frac{B}{u} + Cu \tag{3.25}$$

This equation was derived to describe what happens on the chromatographic column. However, it can be proved that this equation also includes in its constants dispersion effects originating from among the column [16,29]. For higher velocities of the mobile phase the member B/u can be neglected in Equation (3.25) with respect to the fact that $B/u \ll A + Cu$. For these regions of velocity, which are important for rapid analysis by low-dispersion chromatography, the following relationship is obtained by substituting (3.25) by (3.23):

$$\sigma^2_{V,c} = (1 + k)^2 A + \kappa(1 + k)^2 Cu \tag{3.26}$$

If Equation (3.21) is used with substituted Equations (3.6), (3.19), and (3.22), we obtain the relationship for the height equivalent of the theoretical plate:

$$H = \frac{L(\sigma^2_{V,c} + \sigma^2_{V,ex})}{V_R^2} = \frac{L\kappa A}{V_M^2} + \frac{L\kappa Cu}{V_M^2}$$
$$+ \frac{6.4 \times 10^{-3} d_{cap}^6 L_{cap} L u_{cap}}{V_M^2 (1 + k)^2 D_m} \tag{3.27}$$

Equation (3.27) shows that at the higher mobile-phase velocities the importance of dispersion in the capillary from the column decreases proportionally with $(1 + k)^2$. Because $\sigma_{V,c}^2$ is a function of k, with increasing k $\sigma_{V,c}^2$ also increases.

If the constant C from Equation (3.27) is expressed, we obtain the relationship

$$C = \frac{H - A}{u} - 6.4 \times 10^{-3} \frac{d_{cap}^6 L_{cap} u_{cap}}{D_m \varkappa (1 + k)^2 u} \tag{3.28}$$

It is evident from Equation (3.28) that with increasing capacity ratio the value of the constant C in Equation (3.25) decreases.

To verify the influence of extracolumnar spaces in low-dispersion chromatography, we used both analytic columns and microcolumns. The analytic columns had an inside diameter of 4 mm, and they were packed with sorbent of particle sizes 3.2 and 6 μm. The column length was selected so that the theoretically obtainable number of plates was approximately equal in both cases. The microcolumns had inside diameters 0.7 and 0.5 mm, and they were packed with sorbent of particle sizes 5 and 10 μm.

Equation (3.25) was verified experimentally. Coefficients A and C were determined experimentally. Coefficient B was calculated from the relationship $B = 2\gamma D_m/u$, where $D_m = 10^{-5}$ cm^2/s and $\gamma = 0.6$. The correlation coefficients of constant C were in the range of values from 0.9 to 0.7. The dispersion variation of A was about 10%.

For a column packed with particles of $d_P = 6$ μm, an electro-chemical detector with a cell volume of about 7 nl, as well as a spectrophotometric detector with a cell volume of 8 μl, was used. The electrochemical detector values of the theoretical plate height equivalent are given in Table 3.3 and compared with the calculated value:

$$H_{min} = A + 2(BC)^{1/2} \tag{3.29}$$

The experimental value H_{min} is in all cases lower than the calculated value. On the contrary, for the optimal velocity u_{opt} corresponding to H_{min},

$$u_{opt} = \left(\frac{B}{C}\right)^{1/2} \tag{3.30}$$

Table 3.3 Comparison of Experimental and Calculated Values from Equation (3.25)[a]

Solute	k	C (s × 10^3)	C_{theo} (s × 10^3)	ΔC (s × 10^3)	H_{min}^{theo} (μm)	u_{opt}^{theo} (mm/s)	H_{min}^{exp} (μm)	u_{opt}^{exp} (mm/s)
I	0.05	3.77	0.45	3.32	20.30	1.63	18.3	1.0
II	1.0	1.42	1.69	−0.27	21.34	1.02	16.0	1.2
III	2.0	2.07	2.73	−0.66	22.29	0.85	16.0	1.2
IV	3.7	2.41	2.95	−0.54	22.13	0.79	16.03	1.2

[a]Column L = 55 mm, d_c = 4 mm, sorbent Separon SIX C18, d_P = 6 μm. Solutes: resorcinol (I), 4-aminoazobenzene (II), 2-aminoazotoluene (III), N,N-dimethyl-4-aminoazobenzene (IV). Electrochemical detector, cell volume V_c = 7 nl, A = 17.89 μm (mean experimental value) Δ_c = C − C_{theo}.

the experimental value is always higher than that obtained by calculation.

As can be seen from Table 3.3 and Figure 3.4a, coefficient C is theoretically dependent on capacity ratio k. The theoretical values of coefficient C were calculated according to the relationship

$$C_{theor} = E(k) \frac{d_P^2}{D_m}$$

where

$$E(k) = \frac{1 + 6k + 11k^2}{96(1 + k)^2} \tag{3.31}$$

The correspondence of the experimental and theoretical values can be considered as proof that in the arrangement given extracolumnar contributions do not affect the value of the theoretical plate height equivalent. Therefore, this experimental arrangement is suitable for work at higher mobile-phase velocities and consequently also for short analyses. For comparison a column with such parameters was connected to a commercial spectrophotometric detector of cell volume 8 μl. The result of this measurement is also given in Figure 3.4. The coefficient C decreases with increasing capacity ratio.

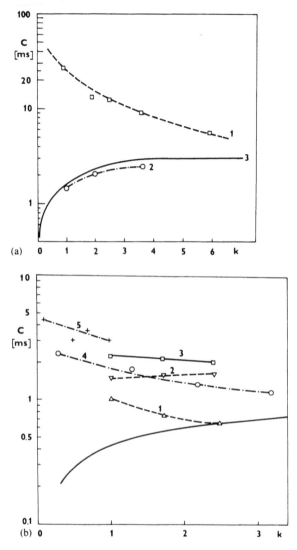

Figure 3.4 (a) Influence of capacity ratio k on coefficient C for particles, $d_p = 6$ μm (Separon SIX C18): (1) UV spectrophotometric detector (cell volume $V_c = 8$ μl); (2) electrochemical detector ($V_c = 7$ nl); (3) theoretical dependence. (b) Influence of capacity ratio k on coefficient C for particles, $d_p = 3.2$ μm. Column $L = 30$ mm, $d_c = 4$ mm, electrochemical detector curve 1, $V_c = 7$ nl, Separon SIX C8; curve 2, $V_c = 7$ nl, Separon SIX C18; curve 3, $V_c = 7$ nl, Separon SIX C18; curve 4, $V_c = 10$ μl, Separon SIX C18; curve 5, $V_c = 20$ μl, Separon SIX C18. Full curve, theoretical dependence.

Table 3.4 Comparison of Experimental and Calculated Values from Equation (3.25)[a]

Solute	k	C (s × 10³)	C_{theo} (s × 10³)	ΔC (s × 10³)	H_{min}^{theo} (μm)	u_{opt}^{theo} (mm/s)	H_{min}^{exp} (μm)	u_{opt}^{exp} (mm/s)
RP 8 (A = 4.68 μm)[b]								
I	0.1	4.38	0.13	4.25	9.80	0.58	9.6	0.4
II	0.5	3.01	0.29	2.72	8.93	0.70	8.5	0.7
III	0.7	3.60	0.34	3.26	9.32	0.65	8.7	0.7
IV	1.0	3.17	0.42	2.75	9.04	0.69	8.7	1.0
RP 18 (A = 6.74 μm)								
I	0.3	2.39	0.21	2.18	10.52	0.79	10.57	2.0
II	1.3	1.79	0.49	1.30	10.02	0.91	11.20	3.0
III	2.2	1.33	0.62	0.71	9.55	1.06	11.20	3.0
IV	3.2	1.17	0.71	0.45	9.39	1.13	11.48	3.0

[a]Column L = 30 mm, d_c = 4 mm. Sorbent Separon SIX C8, 18, d_p = 3.2 μm. For solutes see Table 3.3. Electrochemical detector, V_c = 7 nl.

Evaluation of a column packed with a sorbent of particle size 3.2 μm was carried out in a similar way. We worked with Separon SIX, whose surface was modified with octyl and octadecyl groups. From Table 3.4 a good correspondence between the experimental and calculated values can be seen for both H_{min} and u_{opt}. For coefficient C an opposite dependence on capacity ratio k was found compared to the theoretical dependence; that is, with increasing k coefficient C decreases. As can be seen in Figure 3.4, the character of this dependence is similar to that of the dependence measured with a sorbent of particle diameter d_P = 6 μm and with the spectrophotometric detector.

To verify the relationship (3.28) experimentally, we changed it to a line equation of the form

$$C = M - mx \tag{3.32}$$

where

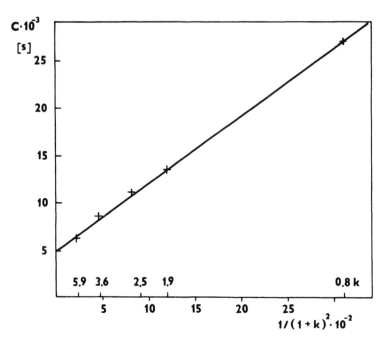

Figure 3.5   Dependence $C = f(k)$ according to Equation (3.32).

$$M = \frac{H - A}{u} \qquad\qquad (3.33)$$

$$x = \frac{1}{(1 + k)^2} \qquad\qquad (3.34)$$

$$m = 6.4 \times 10^{-3} \frac{d_{cap}^6 L_{cap} u_{cap}}{L \varepsilon_u^2 d_c^4 D_m u} \qquad\qquad (3.35)$$

The experimental values C from Figure 3.4 with coordinates $C \simeq x$ are presented in Figure 3.5. The values obtained from a graph correspond to the experimental values. The section on line C in Figure 3.4 and in the area of high k limits the value of $C = 3$ ms. This value corresponds to the values of coefficient C obtained with sufficiently high capacity factors k at which the contribution of extracolumnar

Table 3.5 Values of Quantities for Calculation of
m According to Equation (3.35)

$d_{cap} = 8 \times 10^2$ cm	$L_{cap} = 1$ cm
$L = 5.5$ cm	$d_c = 0.4$ cm
$\varepsilon_u^2 = 0.5$	$D_m = 2 \times 10^{-5}$ cm^2/s
$u_{cap}/u = 2.6 \times 10^{-2}$	

spaces to zone dispersion at higher mobile-phase velocities is negligible. Only under such circumstances can Equation (3.25) be interpreted as an equation of diffusion effects on the column.

The line slope in Figure 3.5 was verified by calculation according to Equation (3.35). The calculation was carried out with the experimental values given in Table 3.5. The value obtained by calculation is $m = 0.5$ ms; the line slope experimentally read from Figure 3.5 is 0.9 ms. This can be considered a satisfactory agreement because the calculation using the values according to Table 3.5 is only an approximation of reality, and the result may be affected by the inaccuracy of some of the quantities (e.g., u_{cap}/u, D_m, and ε_u).

The dependence $C = f(k)$ has been also found in microcolumns [16]. We measured the dependence of the theoretical plate height equivalent on the mobile-phase linear velocity for columns packed with a sorbent of $d_P = 6$ μm, inside diameter $d_c = 0.5$ mm, and length 5, 10, and 20 cm. We also measured columns packed with a sorbent of $d_P = 10$ μm with the same diameter and 5, 10, 20, and 30 cm long. The results are given in Figures 3.6 and 3.7. It is evident from these graphs that the course of the dependence $H = f(u)$ for columns packed with particles of $d_P = 10$ μm is practically the same for all column lengths. The height of the theoretical plate in the minimum of the curve for columns of length from 10 to 30 cm varies from 29 to 32 μm. This result indicates that the member for dispersion in the extracolumnar spaces in Equation (3.27) is negligible and the properties of the column can be described by Equation (3.25). For a column length of 5 cm the extracolumnar space affects the decrease in H_{min} depending on the increasing capacity ratio:

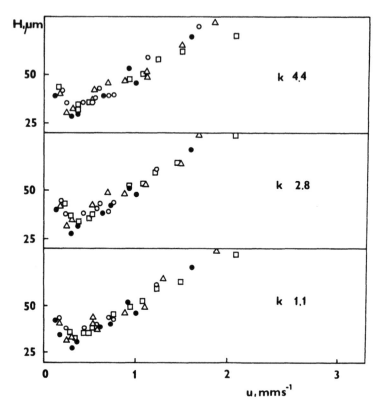

Figure 3.6 Dependence of theoretical plate height equivalent H on mobile-phase linear velocity u for columns packed with particles of $d_p = 10$ μm: (●) $L = 5$ cm; (■) $L = 10$ cm; (△) $L = 20$ cm; (○) $L = 30$ cm.

H_{min} (μm)	k
36	1.1
35	2.8
34	4.4

The contribution of extracolumnar spaces to the theoretical plate height equivalent is noticeable for columns packed with a sorbent of

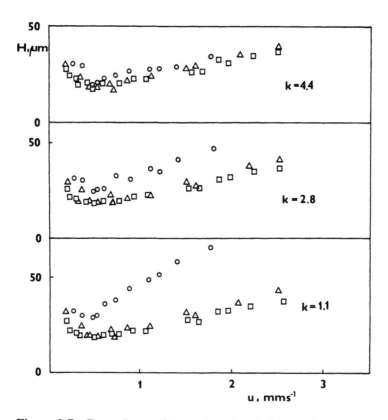

Figure 3.7 Dependence of theoretical plate height equivalent H on mobile-phase linear velocity u for columns packed with particles of d_p = 6 μm: (○) L = 5 cm; (△) L = 10 cm; (□) L = 20 cm.

d_P = 6 μm. Figure 3.7 clearly shows the dependence of the line slope of the linear part of the curves (constant C) on the capacity ratio and the column length. Even if we did not evaluate the values quantitatively, they are in obvious correspondence with Equation (3.27). The technique of packing and injection as well as the electrochemical detector show that microcolumn chromatography can also work with very short columns (L = 3 cm).

3.2.5 Detectors

In liquid chromatography concentration detectors are preferred [51,52], that is, detectors that respond to the solute concentration in the mobile phase. Mass detectors that respond to the mass flow rate of the solute through the detector are limited to some types of transport detectors [53–55], which are usually applied when there are special demands on selectivity [56] or when liquid chromatography is combined with mass spectrometry [57].

The detector cell in microcolumn chromatography forms an extra-columnar space whose shape and especially volume can affect the result of the chromatographic separation. At present microcolumns are used with adjusted detectors based practically on all the principles used in liquid chromatography with conventional analytic columns. The greatest design difficulties were with refractometers [58] and permitivity detectors [59]. The possibility of applying the electrokinetic effect to detection has also been investigated [60,61], especially in packing microcolumns [62] and capillary columns [63]. Despite the indisputable advantages of this detection method—mainly sufficient sensitivity and increasing response to increasing capacity ratios of solutes in the system; the effective volume of the detector correspond-ing to two height equivalents of the plate; and the possibility of solute detection in a selected site on the column—this detection method has not been widely used because of the insufficient stability of the double layer in systems with low conductance. The possibility of increasing the response stability has been further investigated [64,65]. A con-ductometric detector with a sufficiently small measuring cell is also applicable to microcolumn liquid chromatography [45,66,67]. The im-portance of conductometric detectors increases proportionally with the importance of ion chromatography used for rapid, selective, and quan-titative determination, especially of inorganic ions.

Thus far the detectors most often used in microcolumn chroma-tography are spectrophotometric detectors [68,69], fluorometric de-tectors [70], and amperometric detectors [71,73]. It is important to consider the miniaturization of cells of different types of detectors.

So that a certain extent of dispersion of the chromatographic zone beyond the column is not exceeded (e.g., 10%), it is necessary to maintain a relation between the standard deviation of the zone profile caused by the column $\sigma_{V,c}$ ($\sigma_{V^\circ C}$ for a nonsorbed solute) and the standard deviation caused by the extracolumnar contribution, the detector cell volume $\sigma_{V,\text{det}}$:

$$k_1\sigma_{V,c} > \sigma_{V,\text{det}} \qquad (3.36)$$

where k_1 is a constant that depends on the selected allowable distortion; for the previously mentioned 10% this must be less than 0.3.

It is assumed that

$$\sigma_{V,c} = \frac{V_M(1 + k)}{n^{1/2}} \qquad (3.37)$$

and that

$$\sigma_{V,\text{det}} = \frac{V_\text{det}}{12^{1/2}} \qquad (3.38)$$

By combining relationships (3.36) to (3.38) we obtain

$$V_\text{det} = k_1 V_M(1 + k) \left(\frac{12}{n}\right)^{1/2} \qquad (3.39)$$

from which it is evident that as the column volume decreases the volume of the detector cell must also decrease if the highest allowable relative distortion of the chromatographic zone is not to be exceeded.

Imagine a spectrophotometric detector whose signal is proportional to the decrease in the radiant flux that passes through the detection cell. At present these detectors are designed so that the entire radiant flux leaving the measuring cell is trapped in the sensitive area of an optoelectric sensor. The resulting electrical signal is therefore proportional to the loss of the radiant flux ϕ that passes through the cell. According to the Lambert-Beer law the change in the radiant flux is proportional to the cell length, and simultaneously, the total flux is given by the product of the cell cross section and the radiant flux density. It is possible to derive the relationship between the change in the radiant flux and the solute concentration c:

$$-d\phi = \varphi_0 \varepsilon V_\text{det} dc \qquad (3.40)$$

where ϕ_0 is the density of the radiant flux entering the cell, ε the molar absorption coefficient of the solute, and V_{det} the product of the length and the detection cell cross section.

The noise of modern photometric detectors is determined mostly by optoelectric sensors. With decreasing volume of the detection cell the change in the radiant flux with dc also decreases, as does, consequently, the resulting response of the detector. The ratio of signal to detector noise also deteriorates.

For amperometric detectors the dependence between the limit diffusion current of the solute I_{Lim} and the solute concentration c can be simply described by

$$I_{Lim} = k_2 z F c (D_m u_{det} A_2 l)^{1/2} \tag{3.41}$$

derived from the more detailed relationships given in the literature [74,75], where z is the number of exchanged electrons, F is the Faraday constant charge, u_{det} is the mobile-phase linear velocity in the detector cell, A_2 is the electrode surface, l is a characteristic dimension of the cell (for a thin-walled electrode it is the channel width; for a wall jet electrode it is the electrode diameter), and k_2 is a constant that depends on the geometric arrangement of the cell.

The effective volume of the amperometric cell is given by the electrode surface. By balancing the volume flow rates in the cell and in the column the relationship between the mobile-phase linear velocity in the cell u_{det} and in the chromatographic column is obtained:

$$\frac{V_{det} l u_{det}}{A_2} = k_3 \frac{V_M u}{L} \tag{3.42}$$

where k_3 is again a constant that depends on the geometric arrangement of the cell.

By combining relationships (3.41) and (3.42), the dependence of the limit current on the effective volume of the cell is obtained in the form

$$I_{Lim} = k_2 z F c A_2 \left(\frac{V_M u D_M k_3}{V_{det} L} \right)^{1/2} \tag{3.43}$$

This relationship shows that the response, that is, the limit current, is linearly dependent on the electrode surface. The same dependence exists between the electrode surface and the noise so that the relation between the signal and the noise is independent of the electrode surface. A further decrease in the cell effective volume may even improve the relation between the signal and the noise as has been shown [76].

It is clear from a comparison of both types of detection that from the point of view of the obtainable minimum detectable concentration amperometric detectors are more advantageous for microcolumns than spectrophotometric detectors.

At present the lowest operating volumes without losing the necessary sensitivity can be obtained with amperometric detectors. The amperometric detector is not as universal as the spectrophotometric detector because the number of electroactive substances is substantially smaller than the number of substances that absorb the energy of the ultraviolet and visible parts of the spectrum. However, the number of substances determinable by the amperometric detector can be extended by suitable chemical treatment of the solute in a way similar to that used for fluorimetric detectors. It is also important to take into account that spectrophotometers with varying wavelengths mostly do not reach the sensitivity of specially designed photometers working at one wavelength, which are also most often used with packed microcolumns.

Application of the amperometric detector considerably diminishes the extracolumnar spaces, and with the extremely low volume of the detector very good efficiency of the chromatographic columns is obtained and is not distorted by the detector.

3.2.5.1 Optical Detectors

Ultraviolet Photometric Detectors. Photometric detectors are the most often used detectors in liquid chromatography with both conventional columns and microcolumns. To make full use of photometric detectors in microcolumn chromatography, necessary arrangements, mostly of the detector measuring cell and the effluent inlet from the column to the measuring cell, must be carried out. The detection cell volume usually should not exceed 1 μl.

The shape of the detection cell plays an important role not only from the viewpoint of effluent flow but also with respect to application of the cell volume for the photometric signal. The optical path l directly influences the signal magnitude. The absorbance A is expressed by the relationship

$$A = \varepsilon c l \qquad (3.44)$$

where ε is the molar absorption coefficient and c the solute concentration.

Perfect washing of the measuring cell with the effluent is enabled if its shape is cylindrical. According to the direction in which the beam goes through the column, cells are divided into two categories. The first category consists of cells lit along the cylinder axis. These cells are usually of the Z or U type (according to the arrangement of inlets), and their optical path is usually 2–8 mm. The volume ranges from 0.3 to 2 μl. Such a miniaturized cell is shown in Figure 3.8a. Miniaturization of the cell is connected to several contradictory effects. To obtain a sufficiently small cell volume the diameter of the cell cylinder must be reduced. Then, however, the beam leaving the cell illuminates only a small part of the photosensitive element of the detector. The result is increasing noise. On the contrary, a very small cell volume suppresses possible temperature fluctuations and simultaneously decreases the influence of possible flow rate pulses. The result is a decrease in the short-term noise. Many authors also point out the prevailing positive influence of miniaturization on the lessening of noise. It is generally accepted that the minimum detectable concentration in the detector increases in relation to the decrease in the optical path l less than supposed in relationship (3.44).

This enables us to design measuring cells lit perpendicularly to the longitudinal axis of the detector cell. The optical path of such cells is usually several tenths of a millimeter, and their volume ranges from 10 to 100 nl. The cell can be successfully made of a fused silica capillary designed for column preparation. A layer of plastic is removed from the capillary surface, and the capillary is connected to the column and placed in the detector in the measuring cell space. This sometimes leads to the notion that the arranged photometers can detect the solute directly on the column. An example of a detector lit perpendicularly to the longitudinal axis is given in Figure 3.8b.

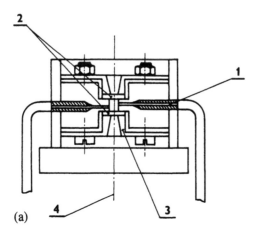

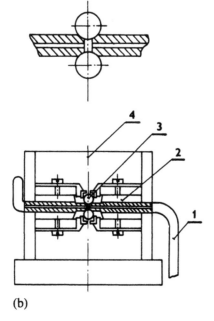

Figure 3.8 Miniaturized photometric cells. (a) Cells lit in the direction of the longitudinal axis of the flow: (1) inlet capillary; (2) quartz plates; (3) plate holder; (4) optical axis of the device. (b) Cells lit perpendicular to the longitudinal axis of the flow: (1) inlet capillary; (2) agate (quartz) balls; (3) ball holder; (4) optical axis of the device.

The analytic properties of spectrophotometric detectors for micro-column chromatography depend not only on the design parameters of the detector cells but, as follows from the principle of spectrophoto-metry, also on the analyzed substance (the molar absorption coeffi-cient). Therefore, the minimum detectable amount of analyte and the minimum detectable concentration can be considered only as orien-tational values. The minimum detectable amount in spectrophotometric detectors for substances with $\varepsilon \simeq 10^3$ cm/mol/L is around 10^{-10} g; the minimum concentration in the detector is usually about 1 μg/L. How-ever, both these values depend on the optical properties of the solute.

As has been mentioned, dispersion of the solute concentration pulse in the detector inlets may cause a significant loss of efficiency of chromatographic separation. In microcolumn chromatography, there-fore, the column outlet must be placed as close to the detector mea-suring cell as possible. In some cases, however, the design of the optical part proper, and sometimes also of the detector electronics, does not allow one to place the column sufficiently close to the detector measuring cell. Therefore, a spectrophotometric detector measuring cell connected by optical fibers to the optical part of the detector, that is, to the radiation source and the optically active element (photoelectric multiplier, photodiode, and the like), was designed [77].

A schematic representation of this cell is given in Figure 3.9. It is evident that the cell can be easily transferred, the connection between the column and the detector is very short, and also the optical path in the measuring cell is extended. The energy losses in the optical fibers are negligible for these fiber lengths ($\simeq 20$ cm). The cell is perfectly thermally insulated from the radiation source, which suppresses the short-term noise of the detector. The detector parameters fulfill well the requirements for the application of microcolumns in liquid chro-matography. For anthranilic acid, for example, the minimum detectable amount of 9×10^{-11} g and the minimum detectable concentration of 2.6×10^{-6} mol/L were determined; from peak dispersion read from the chromatogram a detector volume of 0.04 μl was determined.

It is evident that despite some problems with design detectors based on absorption spectrophotometry remain the most important in liquid chromatography as well as in chromatography with small-diameter

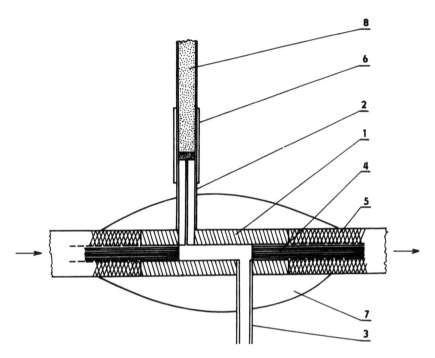

Figure 3.9   Photometric cell with optical fibers: (1) cell body, inside diameter 200 μm, length 1 mm; (2) inlet fused silica capillary of 20 μm ID; (3) outlet capillary; (4) quartz fiber of the light conductor; (5) polyimide cover; (6) connecting capillary; (7) epoxy resin; (8) column of fused silica capillary. Arrows show the beam direction.

columns. Their basic advantage is in the easy selection of detection selectivity over a wide range of detectable analytes, which at present cannot be provided by other types of detectors.

Fluorimetric Detectors.   With respect to their sensitivity fluorimetric detectors belong among the best detectors used in liquid chromatography. The relatively easy design of measuring cells with sufficiently small diameters led to the frequent application of fluorimetric detectors with columns of small diameters.

   If the molecule absorbs photons, it changes into a state of excitation. During transition of the electron to the basic state either a photon is radiated, that is, fluorescence, or the energy absorbed by the molecule changes into heat. The fluorescent flow F is proportional to the solute concentration c:

$$F = 2.303K\phi_0\phi\varepsilon cl \tag{3.45}$$

where K is the efficiency constant of fluorescent effect application, ϕ_0 is the impinging radiation flow, ϕ the quantum efficiency, ε the molar absorption coefficient, and l the optical path length.

Similarly to the spectrophotometric detector, the response of the fluorimetric detector also depends on the molar absorption coefficient and the ability of the solute molecule to fluoresce. The fluorimetric detector response is far from versatile, and therefore the detector is selective. In contrast to detectors based on absorption spectrophotometry in which the response is proportional to the ratio of impinging ϕ_0 and the radiation flow ϕ, the fluorimetric detector response ϕ_f is directly proportional to the impinging radiation flow ϕ_0. Therefore, detectors that use lasers as a radiation source proved successful. The emitted radiation, fluorescence, has a higher wavelength than the excited radiation. The signal is detected by a photosensitive element, most often by a photoelectric multiplier, placed at a suitable angle to the direction of the excitation beam.

The design of the fluorimetric detector measuring cell arises from the requirement for the efficient absorption of the emitted radiation and its transfer to the photoelectric multiplier. Simultaneously the detector noise must be limited as much as possible. Most frequently four basic sources of noise are supposed to exist in fluorimetric detectors, all of them caused by dispersion of excited radiation:

1. Dispersion due to reflection and refraction of radiation on the cell walls
2. Elastic Rayleigh dispersion
3. Raman dispersion
4. Dispersion of radiation on solid impurities in the effluent

Another important source of noise are fluorescent components in the mobile phase. The contribution of individual dispersion types to

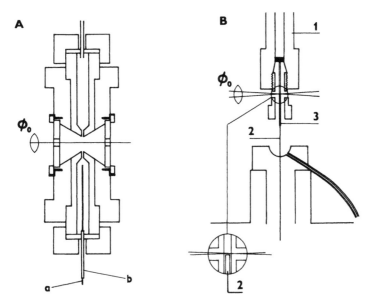

Figure 3.10 Types of fluorimetric cells with laser excitation. (A) Type with auxiliary liquid: a, inlet of effluent; b, auxiliary liquid inlet. (B) Type with radiation discharge by optical fiber. (1) Column; (2) optical fiber; (3) capillary; ϕ_0, incoming radiation. (From Ref. 78.)

the total detector noise depends on the cell design and the material used for its manufacture as well as on the excited radiation flow.

In contrast to spectrophotometric detectors, the measuring cells of fluorimetric detectors determined for work with microcolumns and in some cases also for work with capillary columns are lit perpendicularly to the direction of the flowing effluent. In the literature several types of design have been mentioned; four of them, suitable for combination with the laser [78] as excitation source, are given in Figure 3.10.

The volume of the fluorimetric detector measuring cell can be lowered below 10^{-6} L using a washing liquid (Fig. 3.10A). The effluent from the microcolumn and the washing liquid flow coaxially into the working space of the cell. The effluent from the column therefore forms a liquid cylinder that penetrates the washing liquid that fills the

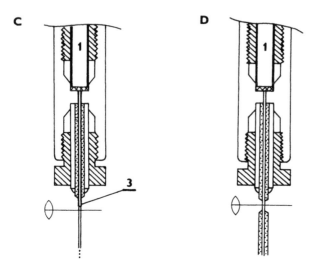

Figure 3.10   Continued. (C) Type with a jet. (D) Type with fused silica capillary.

cell. By selecting the flow rates of the effluent and the washing liquid it is possible to obtain different diameters of the liquid cylinder flowing out of the column. The cell's effective volume is determined by the effluent flow diameter and its lit length. At a suitable flow rate the effective volume equals about 50 nl. A disadvantage of this type of cell is the considerable dispersion of radiation on the walls and windows of a cell whose geometric volume is large. Also, the impurities in a relatively large volume of the washing liquid may increase the detector noise.

   A detector (Fig. 3.10C) has been designed whose cell is formed by the irradiated part of the freely jetting liquid. This method of detection proved successful mainly in connection with conventional columns. The liquid discharge velocity in the jet must not drop below a certain limit value, thus preventing the formation of droplets that would make the detector unusable. It has been proved [78] that the detector can work well at a flow rate above 45 μl/min and jet diameter above

12 μm. The main advantage of this design is a low dispersion of light because there are no fused silica windows and no cell walls.

A number of methods of using fused silica capillaries for the design of fluorimetric cells have been worked out. These consist of an unpacked end of the column which is exposed to excited radiation by removing the plastic cover of the capillary. The fluorescent radiation is sensed at a suitable angle by a photoelectric multiplier (Fig. 3.10D), or this operation is carried out by an optical fiber inserted in the fused silica capillary (Fig. 3.10B). In Figure 3.10B the irradiated volume in the capillary equals 3 nl; nevertheless, the real working volume measured from the liquid outlet from the column to the measured space equals 98 nl.

Fluorimetric detectors with sorbent-packed fused silica capillaries exhibit considerable sensitivity. The minimum detectable amount for coumarin, for example, is 8.4 fg (8.4×10^{-15} g) [79].

Only a limited number of substances exhibit fluorescence, and therefore chemical treatment of the solute beyond the column is often used for the generation of fluorescent compounds. No matter how perfect the design of the reactors may be, at the derivatization beyond the column the peak is always dispersed. Nevertheless, the selectivity and sensitivity of detection are the main advantages of fluorimetric detectors and why they are frequently used in microcolumn liquid chromatography.

Other Optical Detectors Suitable for Microcolumn Liquid Chromatography. Other optical detectors are much less frequently used in microcolumn liquid chromatography because of the difficulty in miniaturizing the detection cell while preserving the corresponding sensitivity parameters, as well as because of the demanding apparatus.

The refractometric detector widely used in liquid chromatography with conventional columns is used in microcolumn chromatography only rarely. The design of the detection cell of sufficiently small volume if often accompanied by considerable difficulty. Simple miniaturization of the cell using a laser as the source of radiation results in diminishing the cell volume below 1 μl. More advantageous seems to be the design of the refractometric detector based on application of the fused silica capillary forming the end of the microcolumn [80]. The laser beam is carried through a microscope body tube to the cover-free section of the fused silica capillary and impinges on a photodiode able to register the deflection of the beam caused by the change in the refractive index in the capillary. The volume of such a cell design based on application of the fused silica capillary column of inside diameter 0.25 mm is below 1 nl. The minimum detectable amount is approximately 100 ng, which corresponds to a concentration in the detector of 30 μg/ml. Even if the sensitivity of refractometric detectors does not reach that of spectrophotometric or even fluorimetric detectors, their main advantage is versatility: they provide a response to every change in the refractive index in the effluent from the chromatographic column. Consequently, this type of detector is unsuitable for work with the mobile-phase gradient. It is most frequently used in connection with the separation and determination of macromolecular substances by gel chromatography.

In detectors determined for microcolumn and in some cases also capillary liquid chromatography, lasers are more often used not only to increase the sensitivity of the miniaturized devices based on classic detection methods (fluorimetry, spectrophotometry, and sometimes refractometry) but also to make use of new methods enabling qualitatively new detection parameters. One of the proposed and, in microcolumn chromatography, successfully applied methods is photothermal refraction [81]. This method of detection uses two beams passing through the detection cell in directions perpendicular to each other. Absorption of radiation from the focused and suitably modulated (90 Hz) beam from the excitation laser (e.g., He/Cd, 3 mW) causes a time-dependent temperature increase in the sample. The subsequent change in the refraction index causes defocus of the measuring beam generated by the other laser (e.g., He/Ne). A simple low surface area photodiode is centered on the profile of the impinging measuring beam and is used for measurement of beam defocus. The measuring cell is formed by a capillary of square cross section 0.2 mm on a side. Interesting detection limits are obtained, for example, 120 fmol acetone, amino acids, and the like.

3.2.5.2   Electrochemical Detectors

The category of electrochemical detectors consists of detection systems that make use of electrical quantities as analytic properties of the effluent. It is therefore possible to detect not only electrochemical reactions on electrodes measured, for example, amperometrically, potentiometrically, or polarographically, but also ionic mobility (conductometry), permittivity, and other quantities. In microcolumn chromatography the most frequently used detectors are the amperometric detectors and, for ion chromatography, conductometric detectors.

Amperometric detectors, as well as, for example, coulometric detectors, are based on electrochemical reaction on the electrode. The detector signal is formed by the flow of electrons released from the working electrode surface. The electrical circuit is closed on the other electrode owing to an electrochemical reaction accompanied by the flow of ions in the mobile phase between both electrodes.

The detector mostly consists of three electrodes (working, reference, and auxiliary) connected to the electrical circuit. In some cases only working and reference electrodes are used. The two-electrode system can exhibit a lower dynamic linear range of response. The material of the working electrode considerably influences detector selectivity and noise. The materials most often used are carbon, mercury, platinum, gold, and copper.

The most important sources of noise are detector electrical elements that cause instability of the potentiostat, the noise of the amplifiers, and the like. Noise can be also caused by variation in some of the properties of the electrodes, such as the changing electrical capacity of the working electrode (due to contamination) or voltage instability of the reference electrode. The noise can be suppressed by perfect screening and earthing of the detector. In some cases the detector must be thermostated.

Figure 3.11 presents six basic types of electrochemical detector cell design [13] that demonstrate that intensive washing of the electrode with effluent must be ensured. The increased velocity of the mass

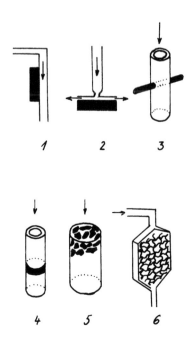

Figure 3.11 Basic design types of electrochemical detectors according to Reference 13: (1) liquid film along a flat electrode; (2) flat electrode placed perpendicular to outflowing liquid (wall jet); (3) wire electrodes in the capillary wall; (4) ring electrode; (5) packing electrode; (6) fiber or fabric electrode.

transfer between the effluent and the electrode surface speeds the process and influences the magnitude of the detector response.

An example of an electrochemical detector [72] with a solid electrode (Pt, Au, and Cu) is provided in Figure 3.12. The cell is made of stainless steel. The effective area of the platinum working electrode is in the center of the electrode head. The effluent from the microcolumn is directed by the detector inlet capillary perpendicularly to the active area of the electrode of diameter $\simeq 0.5$ mm and further flows through the gap between the electrode metal jacket and the openings in the detection cell body. The cell effective volume taken up by the working electrode active area and the conical mouth of the detector capillary tubing is about 4 nl. Results have shown that if the

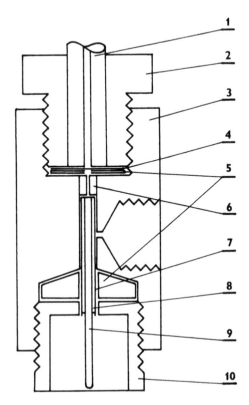

Figure 3.12 Electrochemical detector with solid electrode: (1) column; (2) stainless steel nut; (3) stainless steel detector block; (4) filter paper; (5) PTFE seal; (6) glass capillary (ID 50 μm); (7) stainless steel capillary; (8) PTFE insulation; (9) wire electrode (Pt, Au, and Cu) with diameter 0.5 mm; (10) connector. (From Ref. 72).

cell inside surface is connected as a nonpolarizable electrode, this double-electrode amperometric detector is equivalent in its properties to the triple-electrode detector.

Examples of the minimum detectable concentration obtainable with an electrochemical detector are provided in Table 5.7. Depending on the character of the electrochemical reaction, a sensitivity of up to several nl/L can be obtained. As follows from Figure 3.15 (see later), in some cases precolumn derivatization of the solutes is used with advantage to obtain a high sensitivity.

Properties similar to those of the refractometric detector can be also found with permittivity detector [82]. It has been shown [59] that it is possible to design a cell with a volume of 40 nl provided with two electrodes connected with a Franklin oscillator working at 20 MHz. The detector concentration limit in the best cases is similar to that of the refractometric detector, $\simeq 40$ nl/ml.

The interest in ion chromatography increased the importance of conductometric detection. The conductometric detector is of a very simple design, signal processing is not demanding, it has a wide range of linear response, and with suitable selection of experimental conditions it is extremely sensitive.

With conductometric detection the electrical current I_c between two electrodes, whose magnitude is given by Ohm's law, is measured as

$$I_c = E_c \frac{\rho_c}{C_d} \tag{3.46}$$

where E_c is the voltage on the electrodes, ρ_c the conductivity of the liquid, and C_d the conductometric cell constant.

For the conductance of a liquid in the cell the following general relationship applies:

$$\rho_c = f((\alpha, \lambda, c)_m (\alpha, \lambda, c)_s) \tag{3.47}$$

where α is the degree of dissociation, a the molar conductivity, and c the concentration of the conductive component. The indices m and S correspond to the mobile phase and the solute.

The detector response corresponds to the difference between the conductivity of the mobile phase and the conductivity of the mobile phase with the solute. It seems to be an advantage if the conductivity of the mobile phase $(\alpha, \lambda, c)_m$ is sufficiently low with respect to the conductivity caused by the solute $(\alpha, \lambda, c)_s$. This is obtained by two methods. In the first case part of the detection system is a suppressor column with a ion exchanger on which the highly dissociated components of the mobile phase (e.g., HCl) are transferred to a low-

dissociation substance (e.g., H_2O). This lowers the degree of dissociation of the component in the mobile phase by several orders of
magnitude (in the example given $HCl \rightarrow H_2O$ changes to $\alpha = 1$–
2×10^{-9}), and consequently, the mobile-phase conductivity also decreases. The detector response is then in direct proportion to the solute
concentration at the given degree of dissociation and molar conductivity. In the other method mobile phases with extremely suppressed
dissociations or those whose ion conductivity is considerably different
from the conductivity of the solute are used. In such cases we can
speak about one-column ion chromatography.

One of the advantages of microcolumn liquid chromatography is
the possibility of using a cell of very small volume for conductometric
detection. Two metal electrodes (most often platinum or stainless steel)
are electrically divided from one another by insulation material, usually
glass or PFFE. The insulation forms the cell at the same time. It is
usually a capillary flown by the effluent. The cells described have
volumes of tens to hundreds of nanoliters. The electrodes are provided
with electrical voltage E_c, most often alternating (units of Hz or units
to tens of V). A higher voltage removes problems connected with the
contact resistance of the electrodes. In this case, however, it is necessary to work with cells whose constant C_d is sufficiently high to
prevent an inadequately high current to pass through the cell. The
constant C_d can have a value of approximately 1–10^4 mm^{-1}. The application of direct voltage has also been described [67,83] for conductometric detection with microcolumns.

The minimum detectable concentrations for conductometric detectors are usually around 10^{-7} mol/L; the detection limit with microcolumns is up to 10^{-12} mol. The dynamic linear range is about 10^4.
However, the conductometric detector is nonselective, and its application in a gradient technique is associated with considerable complications, if it is possible at all.

3.2.6 Sample Injection

Injection of the sample in the chromatographic column belongs among
the operations contributing most significantly to dispersion. This is

most evident when working with microcolumns [13] in which dispersion in the injection system may deteriorate the results of the entire analysis.

In the process of injection three main sources of injected sample dispersion can be found. The first is the volume of the injected sample. For an ideal concentration pulse the variation corresponds to Equation (3.15) or (3.16) (also taking into account the column variation and, consequently, the maximum permissible sample dispersion). The second mechanism of dispersion in the injection system is transport of the sample to the chromatographic column. This is the variation originating from sample flow through a capillary of the parameters shown in Equation (3.22). Whereas the first dispersion effect is independent of the liquid flow rate, the second is a function of it. It has been shown [21] that in injection systems currently suitable for work with microcolumns, the second mechanism is most significant. The third source of dispersion in the injection system can be the spaces through which the liquid does not flow directly, so-called traps, from which the liquid passes to the mobile-phase flow by diffusion. These effects originate most frequently in the spaces of the seal between the capillary carrying the liquid from the injector and the column itself. These effects are practically unpredictable and undescribable because the shape and the volume of these spaces are mostly unknown. To prevent significant distortion of chromatographic separation they must be eliminated.

In microcolumn chromatography six-way injection valves and four-way valves with an inside injection loop are most frequently used. It has been shown [21] that the injecting peak is dispersed directly in the inside injection loop during washing out of the sample from the valve by the mobile phase. As mentioned, this dispersion is highly dependent on the mobile-phase flow rate. A further source of dispersion is the connecting capillary between the injection valve and the column. This is the main reason that commercial injection valves are mostly not suitable for injection of the sample in microcolumns if they are not manufactured specially for this purpose. The degree of dispersion in the injection valve is also influenced by the change in flow during injection. The flow change depends on the design of the chromato-

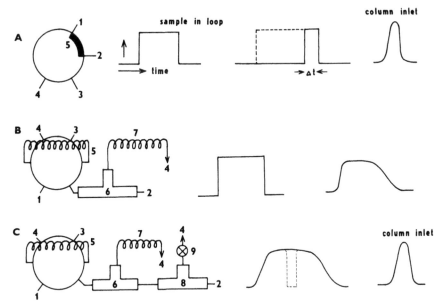

Figure 3.13 Injection systems suitable for microcolumns: (A) valve with internal loop; (B) splitter; (C) pulse-regulated splitter. (1) Inlet of liquid from a pump; (2) inlet to the column; (3) sample inlet; (4) discharge; (5) injection loop; (6) splitter; (7) flow-limiting capillary; (8) discharging splitter; (9) time-controllable stop valve.

graph, that is, on whether the pump works at constant pressure or serves as a source of constant flow.

Several examples of injection systems suitable for microcolumns are given in Figure 3.13A. For small-diameter microcolumns (0.1 mm) in some cases the volume of the sample injected from the four-way valve inside loop is too large. For this case a four-way valve controlled by a time switch was designed [84]. The injection loop is connected to the mobile-phase flow for a time shorter than necessary to wash out the whole loop. The injector works reliably, and the injected volumes are in direct proportion to the time during which the loop is connected to the mobile-phase flow.

Another injection system also successfully used for columns in both gas and liquid chromatography is a splitter. The six-way and sometimes also the four-way injection valve is connected to the splitter. The injected volume is then proportional to the ratio of the flow through branches 5 and 2, represented schematically in Figure 3.13B. The time of sample injection on the column is proportional to the time of its washing out of the injection stopcock loop. This sometimes leads to undesired dispersion of the concentration pulse on the column inlet, as can be seen in Figure 3.13B. This disadvantage is partly eliminated by the system presented in Figure 3.13C [85]. From the concentration pulse originating beyond splitter 6 the greater part is discharged by an automatically regulated control valve and the smaller part is then conducted to the column. This injection system can inject sample volumes of nanoliters with good reproducibility. The injected sample volume depends on the time during which valve 9 remains open at the moment of injection.

Sample injection can also be solved in a new way [24,26]. Dead spaces of the apparatus are reduced as much as possible, and it is possible to influence the mechanism of suction and discharge of the sample. The chromatograph that combines sample injection and pumping of the mobile phase into a single instrumental element is represented schematically in Figure 3.14. It consists of two basic parts: a syringe of volume 100 μl (13) and a liquid distribution block (11). The syringe plunger is operated by a stepper. A needle (12) connected to the syringe enters the liquid distribution block (11) and passes into it through two seals. One (10) seals the column and the other (2) contains inlets to the mobile phases and the sample (1). A side opening in the needle (diameter 0.15 mm) can be adjusted for connection with some of the inlets (1), or it leads into the column through a connector (0.15 × 2 mm). Either thick-walled glass columns, for example $d_c \simeq 0.7$ mm and 30 mm long, or packed fused silica capillary columns are used. The mobile phase flows out of the column through a connector (0.15 × 3 mm) and washes the cross section of the platinum plate of diameter 0.5 mm that functions as the working electrode of an amperometric detector. The detector volume, including the tubing, is less than 60 nl.

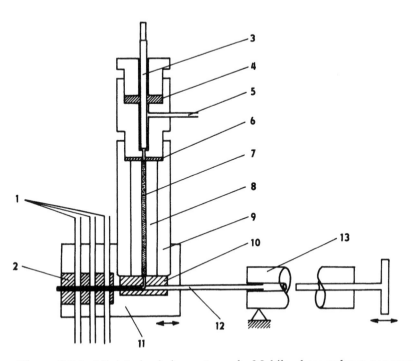

Figure 3.14 Miniaturized chromatograph. Mobile-phase volume per pump stroke is 100 µl: (1) mobile-phase and sample inlets; (2) inlets seal; (3) amperometric detector electrode; (4) silicone rubber seal; (5) effluent discharge; (6) PTFE seal; (7) column packing; (8) glass column, d_c = 0.7 mm, ID 7 mm, length 30 mm; (9) column metal jacket; (10) combined seal of column and injection needle containing a connector of diameter 0.15 mm and 2 mm long; (11) liquid distribution block; (12) injection needle, ID 0.3 mm, OD 0.5 mm, provided with a side outlet of diameter 0.15 mm and closed at the end; (13) syringe of volume 100 µl.

Sample injection is thus connected to the entire chromatographic analysis in a single process. The apparatus functions as follows. The packing opening of the needle (12) is placed under a suitable liquid inlet (1), and the injection needle is filled with the selected mobile phase or with combinations of mobile phases. The opening in the needle is then moved and a sample is sucked into the needle. The given experimental conditions enable one to inject volumes from 0.008 to 2 µl. Volumes less than 0.008 µl cannot be injected with sufficient accuracy; higher volumes of the sample also fill the reservoir of the syringe, which is connected with strong sample dispersion. On the chromatograph described here, of course, volumes exceeding 2 µl, such as 100 µl or even more, can be injected for trace analysis. Under such circumstances, however, it is necessary to make use of enrichment effects and increase the solute concentration at the column inlet (see Sec. 4.3). If the concentration gradient of a component in the mobile phase must be obtained to carry out the analysis successfully, two methods can be used. The analysis begins in an isocratic regime. After pushing a part or the total content of the needle into the column, the mobile phase is sucked in with a higher elution strength and the analysis continues. This method creates a stepwise gradient that can be repeated if necessary. It is also possible to obtain a continuous gradient of mobile-phase composition if small volumes of the mobile phase are gradually sucked into the syringe with higher and lower elution strengths. The content in the syringe is sufficiently dispersed [36,37], and the mobile-phase gradient is formed in the shape of an S. An example of such an analysis is given in Figure 3.15. Similarly, it is possible to focus the concentration pulse of the solute at the chromatographic column inlet.

The degree of dispersion of the concentration pulse during injection was verified experimentally at the direct connection of the electro-chemical detector with the injector. The shape and volume of the concentration pulses were measured under different conditions of injection. The volume between the injection needle side opening and the detector electrode did not exceed 100 nl. The shape of the pulses is evident from Figure 3.16. Volume V_i in which injected volume V_s is transported to the detector is given in Table 3.6. The volume was measured as a point of intersection of tangents to the curve, with the axis representing the volume of the passed mobile phase. The velocity of displacement of the sample corresponds to the mobile-phase flow rate F_m. The degree of injected peak dispersion seems also to be influenced by the velocity of the sample being sucked into the injection needle. With slow suction (0.1 µl/s) volume V_i is always lower than with rapid suction (1 µl/s) under equal experimental conditions. When

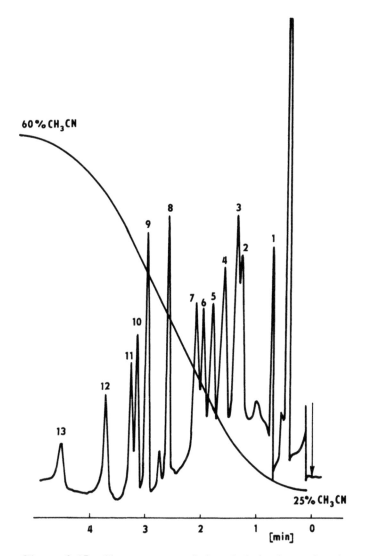

Figure 3.15 Chromatogram of dansyl derivatives of amino acids. Glass column, 0.7 × 30 mm, Silasorb SPH C18, d_P = 7.5 μm; mobile-phase gradient from 25 to 60% CH₃CN in acetate buffer solution (0.01 mol/L), pH = 3.4. Sample: (1) DNS-L-cysteine; (2) DNS-L-asparagine; (3) DNS-L-glutamine; (4) DNS-L-serine; (5) DNS-L-glutamic acid; (6) DNS-hydroxy-L-proline; (7) DNS-L-threonine; (8) DNS-L-α-alanine; (9) DNS-L-proline; (10) DNS-L-enylalanine; (11) DNS-L-leucine; (12) di-DNS-L-lysine; (13) N-O-di-DNS-L-tyrosine (5 × 10⁻¹² mol of each amino acid, V_s = 0.64 μl, amperometric detector, platinum electrode, potential 1.2 V).

Table 3.6 Influence of Velocity of Suction and Displacement of Sample on Peak Dispersion

$F_m{}^a$ (μl/s)	$V_s{}^b$ (μl)	V_i(μl)^c Slow Suction (0.1 μl/s)	Quick Suction (1 μl/s)
0.303	0.27	0.66	1.0
	0.54	1.3	1.45
	1.08	1.7	1.9
0.154	0.27	0.8	0.83
	0.54	1.08	1.23
	1.08	2.1	2.3
0.077	0.27	0.57	0.68
	1.08	1.5	1.77
0.031	0.27	0.47	0.58
	0.54	0.84	1.0
	1.08	1.37	1.78

[a]Velocity with which the sample is pushed on the column, equal to the volume flow rate of the mobile phase through the column.
[b]Sample volume.
[c]Volume in which V_s is transported to the electrochemical detector connected directly to the injector.

evaluating dispersion as the lowest theoretically possible, as follows from Equation (3.15), it is then evident that dispersion obtained at low values of suction and F_m is very low and represents the values $V_i \in \langle 1.1–1.9\ V_s \rangle$. Compared with similar studies carried out earlier [86–88] on chromatographs designed for working with conventional analytic columns, the results obtained are very satisfying. A slight dependence $V_i = f(F_m)$ can be ascribed to the volume of the capillary between the injector and the electrochemical detector electrode. It is assumed that low dispersion of the concentration pulse is also caused by reversal of the liquid flow in the injection needle during injection as a result of which the flow profiles formed at suction are subtracted from the flow profiles formed at the discharge of the sample to the column. This assumption is also supported by the dependence of the

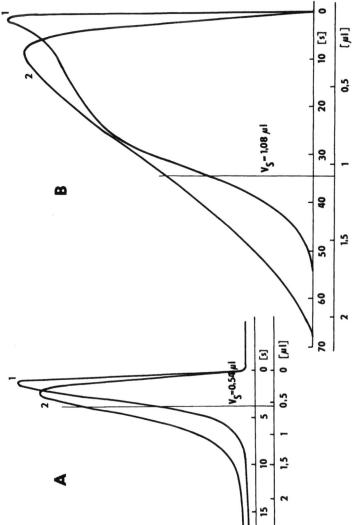

Figure 3.16 Examples of influence of sample suction velocity on peak dispersion. Mobile phase: water; sample: 1% sodium nitrite in water. (A) Injected volume of 0.54 μl; suction velocity: (1) 0.05 μl/s; (2) 0.5 μl/s; displacement 0.15 μl/s. (B) Injected volume of 1.08 μl/s; suction velocity: (1) 0.1 μl/s; (2) 1.0 μl/s; displacement 0.03 μl/s.

shape of the sample concentration pulse and the volume in which the pulse is transported on the column. With slow suction and slow displacement of the sample from the injection needle, the concentration pulse is closer to the ideal rectangular shape at higher flower rates (Fig. 3.16).

The technique of sample injection on the chromatographic column significantly influences the total obtained efficiency of separation. In microcolumn chromatography with very low column diameters (0.2–

Table 3.7 Some Producers of Microcolumn Chromatographs

ABIMED Analysen-Technik GmbH Raiffeisenstraße 3 D-4018 Langenfeld, Germany	Dionex Corp. 4 Albany Park Camberley, Surrey GU 2PL, UK
ABLE Instruments and Accessories A-1011 Vienna, Austria	Fison Instruments/Carlo Erba Bishop Meadow Road Loughborough, Leicestershire LE11
ANSPEC Co Inc. Ann Arbor, MI 48107	ORG, UK
	Genzomed Biotech Inc. 10741 Little Patuxent Parkway Columbia, MD 21044
Applied Biosystems 850 Lincoln Centre Drive Foster City, CA 94404	
	Hewlett-Packard Co 3495 Deer Creek Road Palo Alto, CA 94303-0890
Bischoff Analysentechnik und Geräte GmbH Böblinger Str. 24 D-7250 Leonberg, Germany	Isco Inc. Lincoln, NE 68505
Carlo Erba Strumentazioine SPA Strada Rivoltana 1-20090 Rodano, Italy	JM Science Inc. 4248 Ridge Lea Road Buffalo, NY 14226-1053
Chrompac International BV Knipersweg 6 NL-4330 EA Middelburg, The Netherlands	

Table 3.7   Continued

Knauer GmbH	Philips Analytical
Henchelheimer Str. 9	York Street
D-6380 Bad Homburg, Germany	Cambridge CB1 2PX, UK
Kontron Instruments Spa	Prolabo
Via G. Fantoli 16/15	12 rue Pelie
I-20138 Milan, Italy	F-75011 Paris, France
Laboratory Instruments	Richard Scientific Inc.
Na okraji 325	Novato, CA 94948
Prague 6, Czechoslovakia	
	Shimadzu Scientific Instruments
Life Science Laboratories Ltd.	7102 Riverwood Drive
Sedgewick Road	Columbia, MD 21046
Luton, Bedfordshire LU4 9DT, UK	
	Sonntek Inc.
Lineach Instruments Corp.	Woodcliff Lake, NJ 07675
2325 Robb Drive	
Reno, NV 89523	Spectra-Physics
	Autolab Division
Pharmacia LKB Biotechnology	3333 North First Street
S-75183 Uppsala, Sweden	San Jose, CA 95123

0.7 mm), this question has not been solved fully. This method of injection may enable one to extend the application of these columns. The technique proved to be successful [24], and with columns of very low diameters theoretical efficiencies have been obtained with the electrochemical detector.

3.2.7   Microcolumn Liquid Chromatographs

A number of microcolumn liquid chromatographs are produced at present. They are sometimes called low-dispersion chromatographs, and they most frequently work with columns of inside diameter around 1 mm. Some manufacturers offering chromatographs are listed in Table 3.7.

Manufactured microcolumn liquid chromatographs can be divided into three groups according to their design and application. The largest group is represented by conventional apparatuses equipped with columns of low diameter, mostly 1–2 mm. This type of low-dispersion or small-bore column chromatograph is offered by manufacturers who equip the liquid chromatographs with detectors of sufficiently small volume measuring cells and suitable connections between the injector and the column. The lower flow rate limit of these pumps is 1–5 μl/min. The second category is represented by apparatuses already designed as microcolumn chromatographs. These apparatuses cannot be used for analysis with conventional columns. There is no necessity of compromise between separation efficiency and extracolumnar volumes. Therefore, columns of inside diameter 1–0.2 mm are usually used. The third and at present not very large group of microcolumn chromatographs consists of devices that are not only of microcolumn design but are also determined for a single type of analysis. They work with columns of diameter less than 1 mm. These columns are packed with special materials: sorbents and exchangers for ion-exchange chromatography and gels for exclusion chromatography.

REFERENCES

1. Scott R. P. W.: J. Chromatogr. Sci. 23, 233 (1985).
2. Sagliano N. Jr., Shih-Hsien H., Floyd R. R., Raglione T. V., Hartwick R. A.: J. Chromatogr. Sci. 23, 238 (1985).
3. Rabel F. M.: J. Chromatogr. Sci. 23, 247 (1985).
4. Knox J. H.: J. Chromatogr. Sci. 18, 453 (1980).
5. Novotny M.: J. Chromatogr. Sci. 18, 473 (1980).
6. Reese C. E., Scott R. P. W.: J. Chromatogr. Sci. 18, 479 (1980).
7. Jansen H., Brinkman U. A. T., Frei R. W.: J. Chromatogr. Sci. 23, 279 (1985).
8. Krejčí M., Šlais K., Kouřilová D.: Chem. Listy 78, 469 (1984).

9. Kucera P. (ed.): Microcolumn High-Performance Liquid Chromatography, J. Chromatogr. Library, Vol. 28, Elsevier, Amsterdam, 1984.
10. Novotny M. V., Ishii D. (eds.): Microcolumn Separations, J. Chromatogr. Library, Vol. 30, Elsevier, Amsterdam, 1985.
11. Krejčí M.: Microcolumn and Capillary Liquid Chromatography (in Czech), SNTL, Prague, 1990.
12. Yang F. Y. (ed.): Microbore Column Chromatography, Chromatographic Science Series, Vol. 45, Marcel Dekker, New York, 1989.
13. Scott R. P. W. (ed.): Small-Bore Liquid Chromatography Columns, Chem. Anal., Vol. 72, John Wiley, New York, 1984.
14. Belenkij B. G., Gankina J. S., Malcev V. Q.: Kapilljarnaja židkostnaja chromatografia. Nauka, Leningrad, 1987.
15. Scott R. P. W. (ed.), in: Small Bore Liquid Chromatography Columns. John Wiley, New York, 1984, p. 127.
16. Kouřlilová D., Šlais K., Krejčí M.: Collect. Czech. Chem. Commun. 49, 764 (1984).
17. Verzele M., Le Weerdt M., Dewaele C., DeJong G. J.: LC-GC Magazin 4, 1162 (1986).
18. Scott R. P. W., Kucera P.: J. Chromatogr. 169, 51 (1979).
19. Yang F. J.: J. Chromatogr. 236, 265 (1982).
20. Kucera P.: J. Chromatogr. 198, 93 (1980).
21. Šlais K., Kouřilová D.: J. Chromatogr. 258, 57 (1983).
22. Kouřilóv D., Šlais K., Krejčí M.: Chromatographia 19, 297 (1984).
23. Krejčí M., Šlais K., Kunath A.: Chromatographia, 22, 311 (1986).
24. Krejčí M., Kahle V.: J. Chromatogr. 392, 133 (1987).
25. Vespalcová M., Šlais K., Kouřilová D., Krejčí M.: Česk. Farm. 33, 287 (1984).
26. Krejčí M., Kahle V.: PV 8421-85.
27. Jandera P., Churáček J.: Gradient Elution in Column Liquid Chromatography, J. Chromatogr. Library, Vol. 31, Elsevier, Amsterdam, 1985.
28. Dewaele C., Verzele M.: J. Chromatogr. 282, 341 (1983).
29. Kahle V., Krejčí M.: J. Chromatogr. 321, 69 (1985).
30. Unger K. K., Messer W., Krebs K. I.: J. Chromatogr. 282, 341 (1983).
31. DiCesare J. L., Dong M. W., Atwood J. G.: J. Chromatogr. 217, 369 (1981).
32. Cooke N. H. C., Archer B. G., Olsen K., Berick A.: Anal. Chem. 54, 2277 (1982).
33. Erni F.: J. Chromatogr. 282, 371 (1983).
34. Mellor N.: Chromatographia 15, 359 (1982).

35. Komers R., Krejčí M., in: Laboratorní chromatografické metody (Mikeš O., ed.). SNTL, Praha, 1980.
36. Šlais K., Preussler V.: J. High Resol. Chromatogr. Chromatogr. Commun. 10, 81 (1987).
37. Šlais K., Frei R. W.: Anal. Chem. 59, 376 (1987).
38. Conder J. R., Young C. L.: Physicochemical Measurement by Gs Chromatography. John Wiley, Chichester, 1979.
39. Takeuchi T., Ishii D.: J. Chromatogr. 190, 150 (1980).
40. Takeuchi T., Ishii D.: J. Chromatogr. 213, 25 (1981).
41. Tesařik K., Kaláb P.: Chem. Listy 69, 1978 (1975).
42. Snyder L. R., Kirkland J. H.: Introduction to Modern Liquid Chromatography, John Wiley, New York, 1979.
43. Liso J. C., Ponzo J. L.: J. Chromatogr. Sci 20, 14 (1982).
44. Kahle V.: Chem. Listy, 78, 760 (1984).
45. Šlais K.: J. Chromatogr., 436, 413 (1988).
46. Katz E., Ogan K., Scott R. P. W.: J. Chromatogr. 260, 277 (1983).
47. Katz E., Ogan K., Scott R. P. W.: J. Chromatogr. 270, 51, (1983).
48. Giddings J. C.: Dynamics of Chromatography, in: Principles and Theory, Marcel Dekker, New York, 1965.
49. Horvath C., Lin H.: J. Chromatogr. 126, 401 (1976).
50. Van Deemter J. J., in: Gas Chromatography 1960 (Scott R. P. W., ed.). Butterworths, London, 1960, p. 194.
51. Jinno K.: J. High Resol. Chromatogr. Chromatogr. Commun. 7, 66 (1984).
52. Novák J.: Kvantitativní analýza kolonovou chromatograffi, Academia, Praha, 1981.
53. Scott R. P. W., Scott C. G., Hess J. Jr.: Chromatographia 9, 395 (1974).
54. Krejčí M., Tesařik K.: AO 167.052, U.S. Patent 4,014.793.
55. Stolyhwo A., Colin H., Guiochon G.: J. Chromatogr. 256, 1 (1983).
56. Šlais K., Krejčí M.: J. Chromatogr. 81, 181 (1974).
57. Mc Fadden W. H., Schwartz J. L.: J. Chromatogr. 122, 386 (1976).
58. Imasaka T., in: Microcolumn Separations (Novotny M. V., Ishii D., eds.). J. Chromatogr. Library, Vol. 30, Elsevier, Amsterdam, 1985, p. 159.
59. Hosseiny A., Benmakroha F., Adler J. F.: Anal. Chim. Acta 174, 245 (1985).
60. Šlais K.: Dissertation, Institute of Analytical Chemistry, Czechoslovak Academy of Sciences, Brno, 1978.

61. Šlais K., Krejčí M.: J. Chromatogr. 148, 99 (1978).
62. Krejčí M., Kouřilová D., Vespalec R.: J. Chromatogr. 219, 61 (1981).
63. Krejčí M., Šlais K., Tesařík K.: J. Chromatogr. 149, 645 (1978).
64. Vespalec R.: J. Chromatogr. 210, 11 (1981).
65. Neča J.: Dissertation, Institute of Analytical Chemistry, Czechoslovak Academy of Sciences, Brno, 1986.
66. Tesařik K., Kaláb P.: J. Chromatogr. 78, 357 (1973).
67. Kouřilová D., Šlais K., Krejčí M.: Collect. Czech. Chem. Commun. 48, 1129 (1983).
68. Ishii D., Asai K., Hibi K., Takeuchi T., Nakanishi T.: J. High Resol. Chromatogr. Chromatogr. Commun. 2, 371 (1979).
69. Yang F. J.: J. High Resol. Chromatogr. Chromatogr. Commun 4, 83 (1981).
70. Hirata Y., Novotny M.: J. Chromatogr. 186, 521 (1979).
71. Šlais K., Krejčí M.: J. Chromatogr. 235, 21 (1982).
72. Šlais K., Kouřilová D.: Chromatographia 16, 265 (1982).
73. Goto M., Koyanagi Y., Ishii D.: J. Chromatogr. 208, 261 (1981).
74. Štulík K., Pacáková V.: Chem. Listy 73, 795 (1979).
75. Štulík K., Pacáková V.: Česk. Farm. 30, 241 (1981).
76. Weber S. G., Purdy W. C.: Anal. Chim. Acta 100, 531 (1978).
77. Janeček M., Kahle V., Krejčí M.: J. Chromatogr., 438, 409 (1988).
78. Folestad S., Galle B., Josefsson B.: J. Chromatogr. Sci. 23, 273 (1985).
79. Gluckman J., Shelly D., Novotny M.: J. Chromatogr. 317, 443 (1984).
80. Bornhop D. J., Nolan T. G., Dovichi N. S.: J. Chromatogr. 384, 181 (1987).
81. Nolan G. T., Bornhop D. J., Dovichi N. S.: J. Chromatogr. 384, 189 (1987).
82. Vespalec R., Hána K.: J. Chromatogr. 65, 53 (1972).
83. Kouřilová D., Phuong Thao N. T., Krejčí M.: Int. J. Environ. Anal. Chem. 31, 183 (1987).
84. Coq B., Gretier G., Rocca J. L., Parhault M.: J. Chromatogr. Sci. 19, 1 (1981).
85. McGuffin V. L., Novotny M.: Anal. Chem. 55, 580 (1983).
86. Kirkland J. J., Yau W. W., Stoklasa H. J., Dilks C. D.: J. Chromatogr. Sci. 15, 303 (1977).
87. Golay M. J. E., Atwood J. G.: J. Chromatogr. 186, 353 (1979).
88. Atwood J. G., Golay M. J. E.: J. Chromatogr. 218, 97 (1981).

4
Trace Analysis by Microcolumn Liquid Chromatography

In chromatographic measurement used for trace analysis the trace mass (or trace concentration) of the analyte sample to should emit a signal at the column outlet strong enough to overcome the background noise of the detector. This can be accomplished in two ways: (1) by improvement in the function of the detector, that is, a sufficient decrease in the minimum detectable concentration or mass in the detector, or (2) by enriching the sample with the analyte. Often this can be accomplished only by application of highly sensitive and selective detectors after enrichment. Our attention here is concentrated on enrichment techniques in microcolumn liquid chromatography.

4.1 TRACE CONCENTRATION OF THE ANALYTE

In chromatography sample enrichment by the analyte [1,2] is most frequently carried out using gradient techniques. In gas chromatography the temperature gradient is used, in liquid chromatography mostly the mobile-phase strength gradient, even if it is possible to change the strength of the system retention by changing the temperature [3]. In

94

both cases the retention strength of the chromatographic system is expressed by analyte capacity ratios. The increase in capacity ratios corresponds to the increase in the retention strength of the chromatographic system.

In isocratic techniques a sample of volume V_S is pulse injected into the column, even though the actual volume V_i entering the column is diluted in the injection device. Separation of individual components of the analyzed mixture decreases the original concentration of the analyte in the sample c_o to the concentration of the same analyte in the outlet from the column c_{max} [see Eq. (3.1)]. In terms of the detector response the value c_{max} is the most important quantity. As stated in Section 3.1, $0 < c_{max}/c_o < 1$ and $c_{max} < c_o$. At the column inlet the volume of the injected sample decreases by the amount of the analyte sorbed by the stationary phase. At the column outlet the volume of the components discharged again increases by the desorbed analyte.

To increase the value c_{max} we must apply gradient techniques with which a significant change in the sample inlet volume is obtained as a result of analyte sorption at the beginning of the column. In agreement with Equation (3.2), it is possible to write

$$Q = c_o V_S = c_j V_j \tag{4.1}$$

where c_j is the analyte concentration at the beginning of the column in the mobile-phase volume V_j.

For the isocratic regime the relationships $c_j < c_o$ and $V_j \geq V_S$ apply. For the conditions of gradient elution the relationships $c_j \lesssim c_o$ and $V_j \lesssim V_S$ apply. It is evident that the volume V_j for a given solute is a function of the chromatographic system retention strength. If the system retention strength changes depending on a change in the modifier concentration c_m, the volume V_j can be expressed as a function of the injected sample volume V_S by the solute capacity ratio k:

$$V_j = \frac{V_S}{1 + k(c_m)} \tag{4.2}$$

The functional dependence of the modifier capacity ratio on the concentration $k(c_m)$ is reflected by the modifier isotherm on a given sorbent in terms of system retention strength, with increasing modifier concentration the solute capacity ratio either decreases or increases.

Gradient chromatography is known by a focus effect occurring because the solute migration velocity U, given by relationship

$$U = \frac{u}{1 + k(c_m)}$$ (4.3)

corresponds to mobile-phase composition and, consequently, also to the migration velocity of the modifier concentration $U(c_m)$. The graphic representation in Figure 4.1 presents the ways in which the properties of the modifier influence the focusing mechanism. In Figure 4.1A the increasing modifier concentration focuses the peak because the retention strength of the chromatographic system also increases. The migration velocity U_2 of solute fraction 2 from the region of higher modifier concentration is lower than the migration velocity U_1 of solute fraction 1 from the region of lower modifier concentration. This happens when the modifier is formed by the mobile-phase component with lower elution strength. It has been shown that it also occurs if the trace component is injected in a solvent with a very low elution strength [4]. Figure 4.1B represents the system in which with increasing modifier concentration the retention strength of the chromatographic system decreases. In this case peak focus occurs during the decrease in the modifier concentration in the mobile phase. The possibilities for application of this type of focus in microcolumn liquid chromatography have been demonstrated [5–8].

In microcolumn liquid chromatography the chromatographic system retention strength gradient can be generated by injection of a suitable modifier together with the sample [6–8]. This technique enables us to generate successive gradients [9] of the types shown in Figure 4.1A and B in one experiment without special demands on the apparatus.

Enrichment techniques in gas chromatography are, except for some cases using capillary columns, mostly performed on precolumns. In liquid chromatography enrichment processes are successfully per-

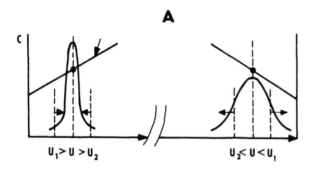

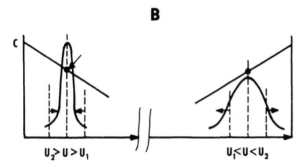

Figure 4.1 Mechanism of focus and dispersion depending on modifier concentration in the mobile phase: c, concentration; U, migration velocity of solute fraction.

formed directly on the analytic column owing to the low values of diffusion coefficients in the mobile phase [1,10,11]. The advantages and analytic importance of enrichment techniques have been shown directly on microcolumns [3,4,6–9,12].

4.2   PEAK FOCUSING TECHNIQUES

Equations (3.15), (3.17), and (3.18) specify the conditions of selectivity under which enrichment techniques can be used. To optimize the conditions for enrichment techniques it is also necessary to pay attention to the dynamics of the injection process [4]. If the symbol $\sigma_{V,c}^2$ according to Equation (3.20) is used again for the solute volume variation on the column and if the volume variance originating on sample injection is expressed as σ_S^2, then in correspondence with Equation (3.14),

$$\sigma_{V,c}^2 > \frac{\sigma_S^2}{k_1} \tag{4.4}$$

where k_1 is a constant that depends on the magnitude of permissible distortion of the peak and also a function of the ratio of the elution strength of the mobile phase and the sample solvent (matrix).

On injection at the six-way stopcock with injection loop of capillary diameter d_1 and volume V_S,

$$\frac{\sigma_S^2}{k_1} = \frac{V_S^2}{k_2\,12} + \frac{d_1^2\,V_S F_m}{k_3\,96 D_m} \tag{4.5}$$

where the first term on the right side of the equation represents the variation in the injected solute volume according to Equation (3.15) and the second term represents dispersion when the injection loop is washed. After rearrangement this term corresponds to Equation (3.22). The constants k_2 and k_3 are again a function of the ratio of the strengths of the mobile phase and the sample matrix. The first term on the right of Equation (4.5) includes the assumption that the sample and the mobile phase are perfectly miscible and that the diffusion coefficients in the mobile phase are identical to those in the sample matrix. It is also assumed that when sample is injected into the mobile phase the elution strengths of the mobile phase and the matrix are identical and $k_1 = k_2 = k_3 = 1$.

To optimize the diameter of the capillary forming the injection loop for a given sample volume V_S, the second term on the right side of Equation (4.5) must be lower than the first; that is,

$$\frac{V_S^2}{k_2\,12} > \frac{d_1^2 V_S F_m}{k_3\,96 D_m} \tag{4.6}$$

assuming that the linear velocity corresponding to the flow rate D_m is higher, then this corresponds to the minimum u_{min} in the dependence of the theoretical plate height equivalent on the velocity. If Equation (3.20) is used for the variation generated on the column, then after rearrangement we may write

$$\sigma_{V,c}^2 < \frac{V_M}{V_S} d_P^2 F_m f(k) \tag{4.7}$$

By combining Equations (4.5) through (4.7) the following inequality may be written for diameter d_1 of the capillary forming the injection loop of the six-way stopcock:

$$d_1^2 < \frac{V_M}{V_S} d_P^2 g(k) \tag{4.8}$$

It is evident from Equation (4.8) that the lower the particle diameter of the sorbent used as microcolumn packing d_P and the higher the injected sample volume V_S, the lower the diameter of the injection loop capillary must be so that the permissible distortion of the injected concentration pulse is not exceeded. The diameter of the injection loop capillary may remain unchanged provided that the ratio between the column dead volume and the injected volume remains constant. Also note that at increased V_S and constant ratio V_M/V_S the loop must be extended. To preserve column efficiency and the corresponding mobile-phase velocity it is necessary to apply greater pressure at the injection loop inlet.

The practical consequences of relationship (4.8) are shown in Table 4.1, which compares two columns with equal separation efficiencies that work under equal experimental conditions, that is, with equal numbers of theoretical plates n, retention times t_R, length L, sorbent particle diameter d_P, linear velocity u, and ratios V_S/V_M. A sample volume $V_S = 1$ ml was selected for the microcolumn; for the column with inside diameter 4 mm the corresponding volume V_S = 32.4 ml. This demonstrates that the same trace analysis results, using detectors of equal parameters, can be obtained with microcolumns that use substantially lower sample volumes than those used with analytic columns. As follows from Table 4.1, the injection loop for the microcolumn can be washed out in 33 s at a pressure of 0.1

Table 4.1 Comparison of Focus Effect at Injection on a Column 150 mm Long and 0.7 and 4 mm ID

Column (mm)	$V_S/$ V_M	V_S (ml)	d_1 (mm)	t_S^a (s)	$\dfrac{d_1^2}{V_S/V_M d_p^2} g(k)$
150 × 0.7, $V_M = 43\ \mu l$	23	1.0	0.5	33	1
150 × 4, $V_M = 1{,}410\ \mu l$	23	32.4	0.5	34,600	1
	23	32.4	1.6	33	10
	4.8	6.8	1.1	33	1

[a]Time necessary for washing the injection loop of volume V_S at pressure 0.1 MPa and viscosity 0.001 Pa · s.

MPa, a pressure easily obtained by a current syringe. For the analytic column the pressure must be increased substantially if the loop is to be washed out in a comparable amount of time. This means that another pump must be used. If the diameter of the capillary forming the injection loop increases, the sample injected on the column is dispersed. To preserve an equal extent of dispersion the ratio V_S/V_M would have to be decreased, which would cause an increase in the minimum detectable concentration, that is, a decrease in the sensitivity of the analysis.

The possibility of injecting a relatively high sample volume into the microcolumn has been verified. An example for chlorinated phenols is given in Figure 4.2. We also analyzed aqueous solutions of polynuclear aromatic hydrocarbons injected through a loop of volume 0.1 ml on a column of $d_c = 0.7$ mm and $L = 150$ mm. The theoretical plate height equivalents H we obtained were perylene, 27 μm, 1,2-benzopyrene, 25 μm, and 20-methylcholanthrene, 28 μm; the theoretical plate height equivalents from a sample of 1 ml volume under equal conditions were 29, 31, 28 μm. The column efficiency remained unchanged; however, as a result of the increase in the injected volume, the duration of the analysis was prolonged. The column works at a constant mobile-phase velocity, and the retention volume increases correspondingly according to Equation (3.18).

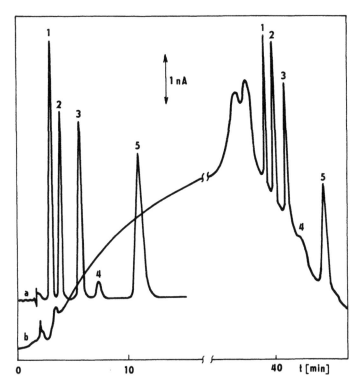

Figure 4.2 Chromatograms of mixtures of chlorinated phenols: (a) 0.2 μl mixture injected in mobile phase, (b) 1 ml aqueous solution injected. Column CGC 0.7 × 150 mm packed with Lichrosorb RP 18, d_P = 7 μm. (1) 4-Chlorophenol (a, 4.5 mg/L; b, 1.6 μg/L); (2) 2,4-dichlorophenol (a, 6 mg/L; b, 4 μg/L); (3) 2,4,6-trichlorophenol (a, 9 mg/L; b, 6.0 μg/L); (4) tetra-chlorophenol (a, 3.5 mg/L; b, 1.6 μg/L); (5) pentachlorophenol (a, 34 mg/L; b, 15 μg/L) (concentrations in injected samples are presented). Mobile phase: water-acetonitrile 40:60 (Vol/Vol) + 0.1 M $NaClO_4$ + 0.001 M $HClO_4$. Linear velocity: a, 1.7 mm/s; b, 1.4 mm/s.

Table 4.2 Detection Limits[a]

V (ml)	Mobile Phase	Solute	Detection Limit (ng/L)
0.1	A	Perylene	170
		1,2-Benzopyrene	260
		20-Methylcholanthrene	1,250
1.0	A	Perylene	40
		1,2-Benzopyrene	60
		20-Methylcholanthrene	260
1.0	B	4-Chlorophenol	20
		2,4-Dichlorophenol	50
		2,4,6-Trichlorophenol	80
		Pentachlorophenol	280

[a]Column 150 × 0.7 mm, Lichrosorb RP 18 (d_P = 7 μm).
Mobile phase: A, acetone-water, 75:25 (vol/vol) + 0.1 M NaClO$_4$; B, acetonitrile-water, 60:40 (vol/vol) + 0.1 M NaClO$_4$ + 10^{-3} M HClO$_4$. Amperometric detector EMD-10.

Experimentally obtained detection limits for selected polynuclear aromatic hydrocarbons and chlorinated phenols are given in Table 4.2.

It appears that sample injection in a solvent whose elution strength is sufficiently lower than the mobile-phase elution strength enables us to obtain detection limits of the order of tens or even hundreds of ng/L. The efficiency of the chromatographic column remains unchanged even on injection of volume samples corresponding to approximately 50 times the column dead volume. The application of such an enrichment technique to microcolumns allows us to substantially increase the sensitivity of the analysis (compared with the sensitivity obtained by equivalent techniques on conventional columns of inside diameter 4 mm, for example).

In most trace analyses it is not possible to select a solvent with a lower elution strength than that of the mobile phase used for separating the components of a mixture. However, even in these circumstances the retention strength of the chromatographic system can be influenced by the contribution of a suitable modifier added to the sample. The

modifier and some of the sample components are eluted simultaneously. Elution of the modifier predictably influences the retention strength
of the chromatographic system in a time-limited interval. This effect
can be used not only for regulation of solutes retention, as in gradient
chromatography, but also for the enrichment process. This procedure
is called an injection-generated gradient.

The ionic compounds separated by ion-pair chromatography are
presented as an example of the injection-generated gradient. In reversed-phase chromatography the ionic compounds often have very
low retentions even if the analysis is carried out with very weak mobile
phases. A suitable modifier—a counterion—can be added to the sample
and by joint injection a dynamic ion exchanger whose retention strength
varies with time is formed. A high concentration of the counterion at
the beginning of the column and its distribution between the stationary
and mobile phases according to the adsorption isotherm cause strong
sorption of the analyzed ionic substances; the retention strength of the
system increases, and consequently, the solutes gather at the beginning
of the column. This corresponds to Figure 4.1B. To enable focus the
peaks must be eluted in the descending part of the gradient [8,13].
Such a situation of course occurs on the column automatically because
the counterion is eluted from the column and its concentration decreases
with time. The character of the decrease depends on the form of the
counterion adsorption isotherm in a given mobile phase and on a given
sorbent [14].

It is known that a mathematical description of the counterion adsorption isotherm on the reversed stationary phase in the mobile phase
is considerably difficult. In some cases Langmuir's isotherm suits the
experiment [15,16]; in other cases Freundlich's isotherm [17,18],
sometimes an exponential type, is proposed for the decrease in concentration of the component eluted in the mobile phase [19]. In our
case the experiment was well described with Langmuir's isotherm [8]
in a narrow range of counterion concentrations in the mobile phase;
in the same system for a wider range of concentrations [7,20] the
results were partly describable by Freundlich's isotherm, partly by its
exponential variation. It is necessary, of course, to know the counterion
phase equilibrium because the capacity ratios of the solutes k depend

on the counterion concentration P in the stationary phase. Equilibrium between the solute and the counterion in the stationary phase is characterized by the distribution constant K_{IE}.

To express the capacity ratios of the solute k, the following equation was used to describe the functional dependence of k on the counterion concentration in the stationary phase $[PC]_s$, where C is used for a coion bounded electrostatically to the counterion:

$$k = \psi K_{IE}[PC]_s + k_R \qquad (4.9)$$

where $\psi = m_s\, c_m V_M$; m_s is the mass of the sorbent in the column, V_M the column dead volume, c_m the coion concentration in the mobile phase, and k_R the capacity ratio of the given solute measured on the reversed phase without using the counterion.

For the functional dependence $[PC]_s = f(P_m)$ Langmuir's isotherm was used [8]. An example of the dependence of the capacity ratios k on the counterion concentration (for naphthalene sodium sulfonane) $[P_m]$ under conditions of isocratic chromatography of ion pairs of catecholamines is given in Figure 4.3. The solid lines represent the results according to Equation (4.9). The capacity ratios of epinephrine measured by injection-generated gradient are represented by the dashed line. From the value of about 1.8×10^{-3} mol/L of the counterion the capacity ratios of epinephrine are constant, independently of the contribution of the counterion to the sample.

If the ionized solute is injected into the column simultaneously with the counterion, the solute moves along the column with variable concentrations of the counterion. Figure 4.4 shows the situation schematically. If a sufficient amount of the counterion is injected into the column simultaneously with the solute, the solute is retained at the beginning of the column. The counterion moves along the column, with respect to the character of the isotherm, as a tailing zone with a sharp front. Simultaneously with the time-dependent decrease in the counterion concentration in the stationary phase the solute capacity ratio also decreases until the moment the solute starts to move along the column.

The situation in the counterion peak tail at the end of the column

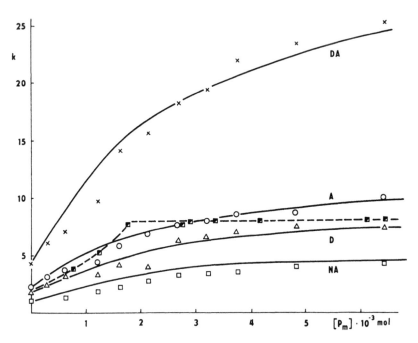

Figure 4.3 Influence of counterion concentration in the mobile phase (P_m) on the magnitude of capacity ratios k. Solid curves correspond to Equation (4.9); the dashed line corresponds to the injection-generated gradient technique. In this case, the value P_m corresponds to the counterion concentration in the counterion peak maximum. Solutes: NA, norepinephrine (noradrenaline); D, DOPA; A, epinephrine (adrenaline); DA, dopamine.

at point L is shown in Figure 4.4. The counterion concentration reaches the maximum $[P_m]_{max}$ at the given corrected retention volume V'_P, and from this moment it gradually decreases. That the solute is eluted at a characteristic counterion concentration $[P_m]_g$ has been derived theoretically [7]. Simultaneously, the focus effect also occurs in agreement with the explanation of Figure 4.1. To obtain this effect by injection-generated gradient it is necessary to add to the sample a certain minimum amount of counterion, as also evident in Figure 4.4. When

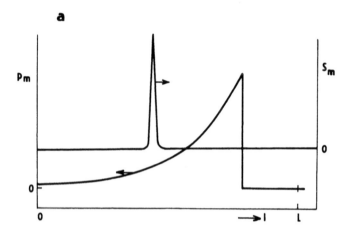

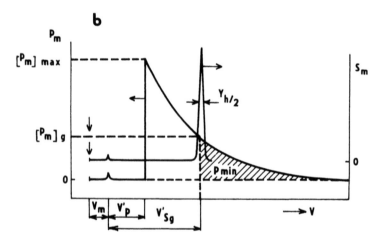

Figure 4.4 Ionized solute elution in counterion zone: (a) Distribution of concentrations of counterion P_m and solute S_m on a column of length L. (b) Concentration of counterion P_m and solute S_m as a function of elution volume V at the column outlet (chromatogram). $[P_m]_{max}$, counterion concentration in peak maximum; $[P_m]_g$, counterion concentration characteristic for solute elution. Hatched area shows minimum amount of counterion P_{min} added to the sample at which the generated retention strength gradient of the chromatographic system is formed by injection. Indices m, P, and S_g designate retention volumes of nonsorbed solute, counterion, or analyzed components eluted in the counterion zone.

evaluating the situation in Figure 4.4, it is evident that both these quantities depend on the character of the sorption isotherm. If we work in the region of Freundlich's isotherm, with the multiplicative constant a and the exponent b, then

$$[P_m]_g = \frac{c_m}{K_{IE}} b(1 - b) \tag{4.10}$$

$$n(P)_{min} = (1 - b)\left(\frac{1}{b^2}\right)^b aS_a m_s [P_m]_g^b \tag{4.11}$$

where S_a is the sorbent specific surface, m_s the sorbent mass, c_m the coion concentration in the mobile phase, and K_{IE} the solute-counterion distribution constant.

Figure 4.5 presents examples of the dependence of reduced retention volumes of the counterion and the solutes on the amount of counterion added to the sample. With sodium o-xylenesulfonane as the counterion (Fig. 4.5), at very small counterion amounts (approximately $P_m = 0.1$ μmol) V_R' is constant: that is, we work in the linear region of the isotherm. At higher counterion concentrations the reduced retention volume decreases. After adding about 10 μmol counterion to 0.1 ml sample the counterion is eluted in the dead volume. The reduced retention volumes of the solutes remain virtually unchanged if the counterion retention volume exceeds the solute retention volume. At the moment the retention volume of the counterion peak maximum decreases below the value of the solute retention volume, the solute retention volume increases and remains constant because the solute is eluted at the counterion characteristic concentration $[P_m]_g$ [Eq. (4.10)] and the condition $n(P) > n(P)_{min}$ is fulfilled. It is evident that from the point of view of retention the measurement results for epinephrine in Figure 4.6 agree with the example in Figure 4.2 even though each measurement was performed under different experimental conditions [7,8].

With regard to both solute peak width and potential solute focus we can see that the solute peak width remains unchanged if the retention volumes of the counterion peak maximum are considerably higher than the solute retention volumes. When the retention volumes of the solute

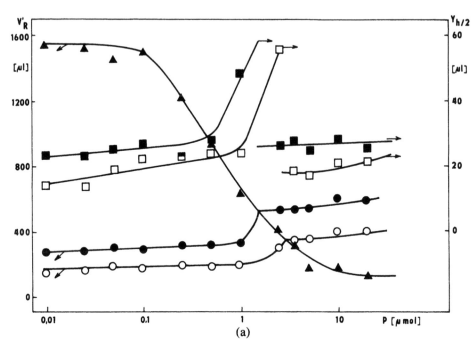

Figure 4.5 Dependence of reduced retention volumes V'_R of counterions (a) sodium o-xylenesulfonane and (b) sodium 1-naphthalenesulfonane and ionic solutes on the amount of counterion P added to the sample. Sample volume 0.1 ml; $Y_{h/2}$, peak width in its middle height. Solutes: filled points, epinephrine; empty points, norepinephrine.

and the counterion begin to come closer (here approximately in the region $n(P) = 1$ μmol), the solute is eluted in the rising part of the counterion peak and the solute peak is dispersed, which is not desired. At the moment when $V'_P < V'_R$ the desired focus occurs; that is, the peak width considerably decreases.

For experimental verification of the analytic applicability of an injection-generated gradient in microcolumn chromatography [5–8] aromatic sulfonic acid salts were selected as counterions so that the course of the time change in the counterion concentration at the column outlet could be recorded spectrophotometrically. The catecholamine chromatogram was recorded by an amperometric detector and is pre-

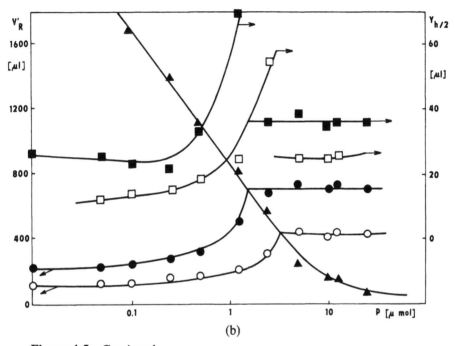

Figure 4.5 Continued

sented in Figure 4.5. The quantitative conditions for the analysis of norepinephrine and epinephrine by injection-generated gradient and isocratic chromatography without using a counterion are given in Table 4.3.

These methods are based on the increase in the retention strength of the chromatographic system at the beginning of the chromatographic column. For peak focus, and for enrichment of solutes on the chromatographic column as well, the reverse effect can also be used, that is, a suitable decrease in the retention strength of the system. We proposed [9] that for this system a modifier be used that exhibits on the given reversed stationary phase a concave isotherm and is consequently eluted from the column in the peak whose front part is diffusively dispersed and whose back part forms a steep front. The shape of the isotherm of substances exhibiting this type of sorption can be easily influenced by the composition of the mobile phase, especially

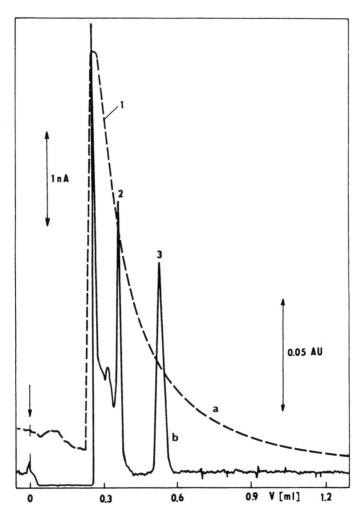

Figure 4.6 Analysis of norepinephrine (2) and epinephrine (3) in the counterion zone (1); dashed line, sodium o-xylenesulfonane; $c = 0.05$ mol/L. Flow rate 2.5 μl/s. EMD-10 amperometric detector (CSFR), Pt electrode 1.0 V (full line). Kratos UV detector, cell volume 0.5 μl (dashed line). Mobile phase 0.1 M $NaClO_4$, 0.001 M $HClO_4$, and 0.001 M EDTS in distilled water.

Table 4.3 Comparison of Chromatographic Analysis of Catecholamines Injected With and Without Counterion[a]

	V_S (μl)	V'_R (μl)	$Y_{h/2}$ (μl)	Minimum Detectable Concentration (mol/L)
Norepinephrine				
Without counterion	1	107	8.2	4×10^{-7}
With counterion	100	375	19.5	3.5×10^{-9}
Epinephrine				
Without counterion	0	200	15.9	9×10^{-7}
With counterion	100	555	28.0	5×10^{-9}

[a]For conditions see Figure 4.6.

its pH. According to the shape of the modifier peak, we may therefore easily control the retention of solutes as well as the extent of peak focus.

We selected methylpyridine as the modifier. Its elution was influenced by the acetic acid concentration in the mobile phase. This example is shown in Figure 4.7. As with the sulfonic acid salt, in this case we also used as modifier a substance that absorbs in the ultraviolet region of the spectrum so that we could spectrophotometrically record its elution from the column. The solutes were detected by an amperometric detector. It is evident from this analysis (Fig. 4.8) that a sufficient extent of selectivity of the amperometric detector allows us to obtain a chromatogram that is not influenced by the presence of modifiers.

We propose the injection-generated gradient to enable both an increase and a decrease in the chromatographic system retention strength.

For trace analysis in microcolumn liquid chromatography the following requirements must be fulfilled:

1. Create conditions for the injection of high sample volumes on the microcolumn so that a high concentration sensitivity can be achieved.
2. Propose the formation of a gradient under the conditions of

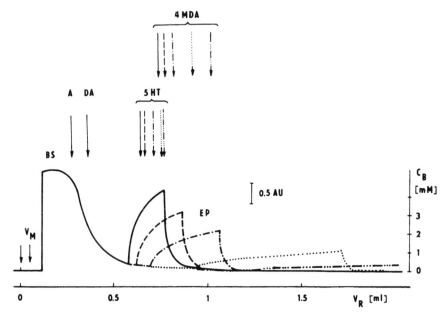

Figure 4.7 Dependence of the zone shape of 4-ethylpyridine (EP) and retention volumes of catecholamines on acetic acid concentration C_B in the mobile phase. Column 150 × 0.7 mm, Separon SIX 18, d_P = 10 μm. Mobile phase 0.2 M Na_2SO_4 in distilled water with acetic acid of concentration (mol/L): 10^{-2} (——), 7.5 × 10^{-3} (- - -), 5 × 10^{-3} (-----), 2 × 10^{-3} (·····), and 1 × 10^{-3} (·-·-·-·). Retention of catecholomines designated by arrows: A, epinephrine (adrenaline); DA, dopamine; 5HT, serotonin; 4MDA, methydopamine. Modifiers (sodium benzenesulfonane, BS; 4-ethylpyridine, EP) detected by Kratos UV spectrophotometer at 211 nm. Catecholamines detected by EMD-10 amperometric detector.

microcolumn liquid chromatography in which miniaturization of the known technical principles in most cases does not lead to the desired results.

3. Stress the influence of the analyzed sample matrix on column selectivity and efficiency.

4. Create conditions for the design of a highly automatable mi-

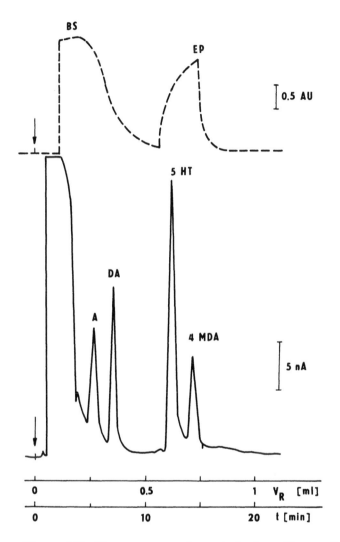

Figure 4.8 Chromatogram of catecholamines. For conditions see Figure 4.7. Acetic acid concentration in mobile phase, 10^{-2} mol/L; sample volume 100 μl, concentration of catecholamines 10^{-6} mol/L; BS 4×10^{-2} mol/L; EP 6×10^{-3} mol/L.

crochromatograph with chemical influence on the chromatographic conditions of the analysis.

These requirements can be also applied to a considerably extent for enrichment precolumns in microcolumn liquid chromatography.

4.3 ENRICHMENT COLUMNS

Focus of analyte peaks on microcolumns represents direct enrichment of the sample by the analyte in the separation column. The techniques and procedures described here are very important in microcolumn chromatography, especially with respect to the decreasing difference between the retention volumes of the components and the lowest volumes of injected sample technically obtainable. For trace analysis proper enrichment of the sample on precolumns is preferred. The significance of precolumns increases because they can fulfill several functions simultaneously. In automation of analytic operations sample treatment before chromatographic analysis is the most time-consuming operation. As discussed subsequently, the automated treatment of samples using precolumns seems to be more efficient and cheaper than using laboratory robots.

The basic advantage of precolumn techniques is the possibility of sorbent selection independently of the required properties of the separation column, which leads to better application of sorbent selectivity for sample treatment. It is also important that on the precolumns either the analyte or the undesired sample components can be sorbed preferentially as required. Another advantage of precolumns is the possibility of application of a sorbent with a larger grain diameter and simultaneously a lower pressure decrease necessary to wash out the precolumn with a liquid. If a sorbent of greater particle diameter is used, there is danger of inadequate dispersion of the sample on its direct transfer to the column. The disadvantage of precolumns compared with enrichment directly in the column is that the procedure is more difficult to automate and in most cases an additional pump is needed to push the sample through the precolumn.

With modern methods of sample treatment the absorption liquid for extraction of the analyte from the liquid or gas sample is substituted in precolumns by sorbents. These sorbents serve in precolumns first to enrich the sample by the analyte and sometimes also to refine the sample of undesired substrates. Sorbents may also enable safe transport or deposition of the sample. Precolumns are connected directly to the microcolumn chromatograph only if they are used for enrichment and have adequate parameters. In other cases plastic precolumns are used in the form of a cartridge or a bed in the syringe, for example. Enrichment precolumns must fulfill the technical requirements of microcolumn liquid chromatography, and precolumns used without a direct connection to the chromatograph usually have an inside diameter of several millimeters and a length of about 1 cm.

The sorption system in enrichment columns should be selected according to the principles described in Section 4.1. In liquid chromatography the equilibration method [21] designed for gas chromatography cannot be used. This method supposes washing out of the column with such an excess of sample that a sorption equilibrium between the analyte and the sorbent is obtained. From the known distribution constant (or capacity ratio) the concentration of the analyte in the sample can be calculated. The analyte distribution constant is influenced by the sample matrix to such an extent that the equilibration method cannot provide satisfactory quantitative results. Therefore, analyte conservation is always used. It is assumed that the total analyte from the known sample volume is retained on the column. The important quantity at the enrichment columns is therefore penetration of the analyte through the column. This quantity can be determined experimentally; however, estimation according to the following relationship is easier and in most cases sufficient:

$$V_S = V_R = V_M(1 + k_s(c))$$ (4.12)

where $k_s(c)$ is the capacity ratio of the analyte on the precolumn sorbent using the matrix as the mobile phase and V_R represents the retention volume equal to the maximum sample volume V_S at which the analyte remains in the column because of penetration of the analyte through the sorbent bed. The capacity ratio $k_s(c)$ is concentration dependent. Because in a suitably selected system the isotherm has the character of the Langmuir or Freundlich isotherm, the capacity ratio increases with decreasing analyte concentration in the sample. This is, of course, an advantage for trace analysis. In practice selection of the sorption system in enrichment columns is not always as difficult as it might seem. For defining the sorption system the (not quite correct) terms "polar" and "nonpolar" are often used. However, in this case these terms are practically always relative properties of the sorbent, the analyte, or the matrix. The sorbents and the sorption systems are given in Tables 4.4 and 4.5. Table 4.5 also shows qualitatively how the capacity ratios k_s decrease from a system consisting of the nonpolar sorbent (e.g., polymeric adsorbent, silica gel with the surface modified by alkylsilane, or carbon adsorbent), the polar matrix (e.g., water), and the nonpolar analyte (e.g., hydrocarbon) to a system in which a sample with a nonpolar matrix (e.g., hydrocarbon) is injected onto a nonpolar adsorbent with the aim of retaining the polar component (e.g., methanol or water). In the first case k_s may reach high values (the enrichment factor in our experiment is $V_S/V_M = 10^4$); in the second case $V_S \doteq V_M$ and consequently $k_s \rightarrow 0$ and no enrichment can be obtained in the given system. Analogously with the decreasing values of k_s, the applicability of the selected chromatographic system for enrichment columns also decreases.

These nonpolar and polar adsorbents (e.g., silica gel and alumina) are nonselective, and on interaction with the analyte their dispersion or electrostatic strengths are most active. The advantage of working with precolumns is the possibility of using selective adsorbents, for example ion exchangers or adsorbents with surfaces modified with metal ions [22]. The analytes are selectively sorbed, and the enrichment factor can be very high. In most cases selective sorptions can be described by a nonlinear isotherm, which enables the analysis of very low concentrations. In an example mentioned in the literature [22], a Cu^{2+} silica gel surface is modified with iminodiacetate on which catecholamines are sorbed as a result of the formation of complexes. The phase equilibrium is considerably influenced by the sample pH. Similar

Table 4.4. Sorbents

Nonpolar	
C18 octadecyl	Si—$C_{18}H_{37}$
C8 octyl	Si—C_8H_{17}
C2 ethyl	Si—C_2H_5
CH cyclohexyl	Si—C_6H_{77}
PH phenyl	Si—C_6H_5
Polar	
CN cyanopropyl	Si—$CH_2CH_2CH_2CN$
2OH diol	Si—$CH_2CH_2CH_2OCH_2CH$—CH_2
	$\quad\quad\quad\quad\quad\quad\quad\quad\quad$ OH $\;$ OH
Si Silica	Si—OH
NH_2 aminopropyl	Si—$CH_2CH_2CH_2NH_2$
PSA N-propylethylenediamine	Si—$CH_2CH_2CH_2N$—$CH_2CH_2NH_2$
	$\quad\quad\quad\quad\quad\quad\quad$ H
NO_2 N-propylnitrophenyl	Si—$CH_2CH_2CH_2$—C_6H_4—NO_2
Ion Exchange	
SCX benzenesulfonylpropyl	Si—$CH_2CH_2CH_2$—C_6H_4—SO_3^-
PRS sulfonylpropyl	Si—$CH_2CH_2CH_2$—SO_3^-
CBA carboxymethyl	Si—CH_2COO—
DEA diethylaminopropyl	Si—$CH_3CH_3CH_2N^+$ $(CH_2CH_3)_2$
	$\quad\quad\quad\quad\quad\quad\quad$ H
SAX trimethylaminopropyl	Si—$CH_2CH_2CH_2N^+$ —$(CH_3)_3$

effects are provided by ion exchangers [23] for enrichment of anorganic cations. The concentration of the complex-forming agent (e.g., tartaric acid) also influences the phase equilibrium.

If columns connected directly to the chromatograph are used, it is necessary to ensure desorption of the analyte in the lowest possible volume. It is an advantage if the desorption agent can be the mobile phase, which is further used in separation on microcolumns. This often occurs

Table 4.5   Sorption Systems for Enrichment Columns

	Sample			Suitability of the
Sorbent	Matrix	Analyte	k_s	System
Nonpolar	Polar	Nonpolar	$\mid k_s$	Suitable
Nonpolar	Polar	Polar	$\mid k_s$	
Nonpolar	Nonpolar	Nonpolar	$\mid k_s$	↓
Nonpolar	Nonpolar	Polar	↓ $k_s \rightarrow 0$	Unsuitable
Polar	Nonpolar	Polar	$\mid k_s$	Suitable
Polar	Nonpolar	Nonpolar	$\mid k_s$	
Polar	Polar	Polar	$\mid k_s$	↓
Polar	Polar	Nonpolar	↓ $k_s \rightarrow 0$	Unsuitable

when the sorbent in the precolumns is of a different polarity than the sorbent in the microcolumn. If the sorbents are identical, the mobile phase desorbs the analyte in a higher volume, which leads to peak dispersion due to the nonideal injection of the solute into the microcolumn. Under such circumstances it is of advantage to desorb the analyte from the pre-column by a liquid with lower elution strength on the microcolumn. The result is peak focus, described in Section 4.1. From the analytic point of view this method of desorption is most useful because it eliminates both the technical imperfections of the arrangements (disproportionally high volumes of the connections in the chromatograph and unsuitable particle diameter of the sorbent in the precolumn) and also a possible variation in the mobile-phase velocity.

This last method of desorption enables us to use precolumns of different diameters. The smallest precolumns are usually made of fused silica capillaries of inside diameter 0.2 mm and several centimeters long. Metal precolumns, as well as thick-walled glass capillaries, usually have an inside diameter of around 1 mm and a length from several millimeters up to several centimeters. All these precolumns are adapted for easy mounting on the chromatograph.

One of the possible connections of precolumns to the chromatograph is presented in Figure 4.9. The sample taken from reservoir 3

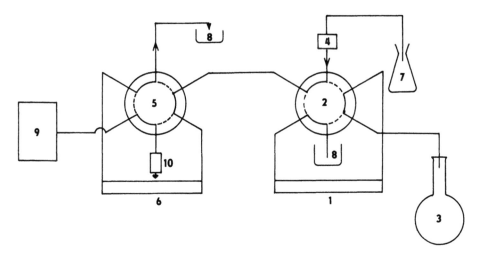

Figure 4.9 Injection device with enrichment column: (1) injection loop; (2) injection six-way stopcock; (3) sample reservoir; (4) auxiliary pump; (5) six-way stopcock; (6) enrichment column; (7) reservoir of washing liquid; (8) discharge (9) mobile-phase pump; (10) microcolumn.

can be measured in injection loop 1 connected to six-way stopcock 2. After switching stopcock 2 to the second position (in Fig. 5.8 marked by the dashed line) the sample is discharged from loop 1 through stopcock 5 to enrichment column 6. As long as a sufficient volume of the liquid from reservoir 7, which can be measured at liquid outlet 8 from stopcock 5, passes through enrichment precolumn 6, stopcock 5 is switched to the position marked in Figure 4.9 by the dashed line. Now the enrichment column is connected to the stream of the mobile phase from pump 9, and the contents of precolumn 6 are carried to microcolumn 10.

In this arrangement there are three liquids, which can exhibit different elution strengths with respect to the analyte and the sorbent used in precolumn 6. The perfect miscibility of all three liquids is a necessary basic condition. It is also of advantage if the washing liquid from reservoir 7 has the lowest elution strength of all three phases. The result is focus of the analyte contained in the effluent from enrichment column 6 on the microcolumn 10.

The arrangement presented in Figure 4.8 has a number of variants. A simple method permits leaving out the system with stopcock 2 and injecting the measured sample directly into enrichment column 6 through stopcock 5. In another arrangement loop 1 is used for the other enrichment column. Suitable selection of washing liquids or selective sorbents enables us to obtain double-step enrichment of the analyte. However, in all the connection variants it is necessary to take into account the elution strength of individual liquids.

With respect to the considerable variability of the problems of trace analysis, it is practically impossible to find an optimum and generally applicable system of enrichment precolumns. For biologic analyses or analyses of environmental components it is usually necessary to remove the matrix from the sample because it forms in the chromatogram a complex tangle of peaks that cannot be distinguished even by a selective detector. Even under such complex conditions, the techniques of enrichment precolumns can be used for determination of concentrations in the range of ppm or ppt, or even lower. In successful analyses the standard deviation does not exceed 10%.

REFERENCES

1. Jandera P., Churáček J.: Gradient Elution in Column Liquid Chromatography. J Chromatogr. Library, Vol. 31, Elsevier, Amsterdam, 1985.
2. Huber J. F. K., Becker R. R.: J. Chromatogr. 142, 765 (1972).
3. Krejčí M., Kouřilová D.: J. Chromatogr. 91, 151 (1974).
4. Šlais K., Kouřilová D., Krejčí M.: J. Chromatogr. 282, 363 (1983).
5. Jinno K.: High Resol. Chromatogr. Chromatogr. Commun. 7, 66 (1984).
6. Krejčí M., Šlais K., Kouřilová D., Vespalcová M.: J. Pharm. Biomed. Anal. 2, 197 (1984).
7. Šlais K., Krejčí M, Kouřilová D.: J. Chromatogr. 352, 179 (1985).
8. Krejčí M., Šlais K., Kunath A.: Chromatographia 22, 311 (1986).
9. Šlais K., Krejčí M., Chmelíková J., Kouřilová D.: J. Chromatogr. 388, 179 (1987).
10. Guinebault P. R., Broquarie M.: J. Chromatogr. 217, 509 (1981).

11. Broquarie M., Guinebault P. R.: J. Liquid Chromatogr. 4, 2039 (1981).
12. Krejčí M., Kahle V.: J. Chromatogr. 392, 133 (1987).
13. Wahlund K. G.: J. Chromatogr. 115, 411 (1975).
14. Conder J. R., Young C. L.: Physicochemical Measurement by Gas Chromatography, John Wiley, Chichester, 1979.
15. Karger B. L., Le Page J. N., Tanaka N., in: High-Performance Liquid Chromatography (C. Horváth, ed.). Vol. i, Academic Press, New York, 1980, p. 113.
16. Goldberg A. P., Navokovská E., Antle P. E., Snyder L. R.: J. Chromatogr. 218, 327 (1981).
17. Deelder R. S., Van den Berg J. H. M.: J. Chromatogr. 218, 327 (1981).
18. Kraak J. C., Jonker K. M., Huber J. F. K.: J. Chromatogr. 142, 671 (1977).
19. Gareil P., Personnaz L., Ferand J. P., Claude M.: J. Chromatogr. 192, 53 (1980).
20. Nevrlá J.: Diploma Thesis, Masaryk University, Brno, 1984.
21. Novák J.: Kvantitativni analýza kolonovou chromatografií, Academia, Praha, 1981.
22. Frei R. W.: Swiss Chem. 6, 11 (1984).
23. Kouřilová D., Phuong Thao N. T., Krejčí M.: Int. J. Environ. Anal. Chem. 31, 183 (1987).

5
Capillary Columns

In 1978 several works [1–6] were published that demonstrated the possibilities of open-tube capillary columns in liquid chromatography, inspiring investigation of a new chromatographic system. Theoretical analyses [7–9] have since proved not only the technical limitations but also the potential uses of capillary liquid chromatography. The effective application of capillary columns was expected to achieve high efficiencies, around 10^6 theoretical plates, and this was also soon obtained in practice [10]. Despite this early success, capillary liquid chromatography has been applied in practice only very gradually because of technical problems that limit the efficiency that can be achieved (mainly detectors and sample injection techniques) and factors that limit the stationary-phase stability and sorption capacity [11].

Several studies have been published that try to avoid the disadvantages of laminar flow in capillary columns and to work with columns of greater diameters. Whether this work was concerned with the turbulent flow [12], with the generation of secondary flow by suitable deformation of the column capillary [4,13], or with segmentation of flow by gas bubbles [14,15], these investigations remained in the realm of speculation and theoretical study. None of these multicapillary systems [16] proved their practicality.

The presumption [7,17] that the conditions in capillaries of very small diameter can also be described by Golay's equation [18] has been proved [17], and gradually considerable attention began to be paid to the preparation of columns with a diameter less than 30 μm [17,19–22].

It is clear that column efficiency is essentially influenced by extracolumnar contributions to chromatographic peak spreading, so that most of these studies investigated spreading of the nonsorbed peak in the chromatographic system. The practical application of capillary liquid chromatography resulted in substantially lower efficiencies than those obtained theoretically. Most of the techniques for capillary column preparation were taken from gas chromatography in which they had been successfully used for a long time. For liquid chromatography [23] various adjustments of the capillary column wall were proposed: adsorption systems, liquid-liquid systems, ion exchangers, chemical modification, surface treatment for reversed chromatography, and cross-linked polymers. Most of these materials proved to be of sufficiently long stability in gas chromatography. The reasons and conditions for the instability of capillary column sorption properties in liquid chromatography have been discussed [11]. To improve the efficiency of capillary chromatographic systems for working with columns of inside diameter from about 5 μm, electrochemical detectors with suitable parameters were designed [24–28].

In most cases glass or fused silica capillary columns are used. Glass capillary columns enable us [29] to observe the column preparation, they can be easily produced on the device used for preparation of capillaries in gas chromatography, and at 60–800 μm in diameter they resist pressures up to 120 MPa [30]. A silica gel layer can be formed on the inside surface of glass capillaries. The essential advantages of fused silica capillaries are chemical homogeneity, inside surface inertness, and mechanical ruggedness. The pressure resistance of the inside diameters of capillary columns can also be considered adequate. The number of applications of fused silica capillary columns in liquid chromatography has also increased more quickly than those of glass columns.

5.1 PUMPS AND INJECTION SYSTEMS

At present pumps specially designed for capillary liquid chromatography are not available on the market because the method is still in the development stage and because of technical demands connected with the pumps. For capillary columns of suitable inside diameter (d_c < 15 μm) the volume flow rates are of the order of μl/min and less, that is, under nl/s. The required pressures under which the pumps should work are approximately 10 MPa. The design of pumps with such parameters that operate as a source of a constant (hard) flow rate is very difficult. Even the smallest volumes shifting between the pump and the column can markedly influence the magnitude of the flow rate.

Most authors who describe experiments with capillary columns use pumps designed for microcolumn liquid chromatography. However, the lower velocity limit of the pumped liquid is then usually higher than is suitable for capillary columns. This situation can be solved in two ways.

The first possibility is to operate the pumps in the regime of constant pressure. Some reciprocal pumps controlled by a microprocessor allow one to adjust and maintain the required pressure. Many cases, especially those in which syringe pumps with a higher working cylinder volume were used, proved that it is useful to adjust the required pressure by the flow of the pumped liquid and then to stop the mobile-phase pumping. The pressure drop due to the flow of the pumped liquid through the capillary column is very small and, therefore, easily controllable. However, both methods assume that the column resistance to the mobile-phase flow does not change with time. This condition is often fulfilled if the pumped liquid is adequately filtered and the mobile-phase gradient changing the viscosity of the pumped liquid is not used.

The other possibility of using a pump with a higher minimum nominal pumped flow rate than required is to include a splitter between the pump and the column (usually between the sample injector and the column). The advantage of this method of pumping the liquid to the capillary column is that the pump can be operated in the regime for which it is designed. The splitter often simultaneously solves the problem of injection of a sample of sufficiently small volume. If a splitter is used, remember that it is necessary to check the constant flow through the column even if the pump is used as the source of the flow. The total flow produced by the pump is divided in inverse proportion to the hydraulic resistance of the column and the dividing branch, and consequently, any change in capillary column resistance or the splitter discharge branch causes a change in flow through the column. The system works as a source of liquid flow at constant pressure in the splitter. In contrast to gas chromatography, in which the needle valve accurately regulates gas flow through the splitter discharge branch, this element is not satisfactory for liquid chromatography. The splitting ratio in liquid chromatography is usually as higher as 1:10,000, and a regulating element with adjustable resistance cannot guarantee that the flow is sufficiently constant. Better results are obtained with packed columns of conventional diameter, which create sufficiently constant resistance in the discharge branch.

A device for formation of a mobile-phase gradient has not yet been described for capillary liquid chromatography. All the techniques originated with the principles described in Section 3.2.2 using a mobile-phase splitter. This situation has undoubtedly limited the importance of capillary liquid chromatography and the speed of its development.

The sample is injected into the capillary column through the splitter. The most common technique is well known from gas chromatography: a simple splitter in which the injected sample volume is divided between the column and the splitter discharge branch in the flow ratio. The possible defects of this technique follow from the character of the liquid flow through the splitter and the constancy of the splitting ratio, and they can be eliminated by techniques of sample injection described in connection with microcolumn liquid chromatography in Section 3.2.6.

5.2 CAPILLARY COLUMNS

5.2.1 Preparation

It is an advantage to prepare capillary columns from glass [30,31]. The capillaries are drawn on the same device used for preparation of capillary columns for gas chromatography, designed by Desty et al.

[32]. Capillary columns with inside diameters suitable for liquid chromatography can be prepared by one of the methods described here [17].

A one-step method is currently used for capillaries with higher inside diameters. The initial glass tube should have an outside diameter of about 8 mm and inside diameter of about 0.3 mm. Under such circumstances columns with an inside diameter of 5 μm and greater can be prepared while preserving the acceptable outside diameter of 0.7–0.8 mm.

A two-step method enables one to prepare suitable capillaries [33] without using the initial thick-walled tube. A currently used tube of 7 mm outside diameter and inside diameter 1 mm is first used for drawing a capillary of outside diameter 0.95 mm, which is then inserted in another tube with the original dimensions, and both tubes are drawn to a capillary with outside diameter from 0.7 to 0.8 mm. The inside diameter of such a combined capillary corresponds to the required dimensions.

The resulting inside diameter for both these methods is determined by the temperature of the drawing device oven. The onset of drawing is given by the temperature of glass softening, and the capillary dimensions after obtaining dynamic equilibrium in the oven correspond to the relationship $d_{c1}:d_{c2} = \sqrt{v_1}:\sqrt{v_2}$, where d_{c1} and d_{c2} are the diameters of the tube or capillary, respectively, and v_1 and v_2 the velocity of placing the tube or capillary in the oven or taking it out of the oven.

With increasing temperature glass begins to overheat; the capillary closes inside, and its inside diameter diminishes. If the temperature increases by about 30 K, a glass rod without an inside hole is produced by the drawing device. The course of capillary narrowing with increasing temperature is evident from Figure 5.1.

The total temperature interval in which the capillary of required diameter can be obtained is 20–30 K (Fig. 5.1). For each kind of glass, of course, the initial temperature and the range of the temperature interval are different. Therefore, these should be verified first for a given kind of glass.

At present, fused silica capillaries are used most frequently. They

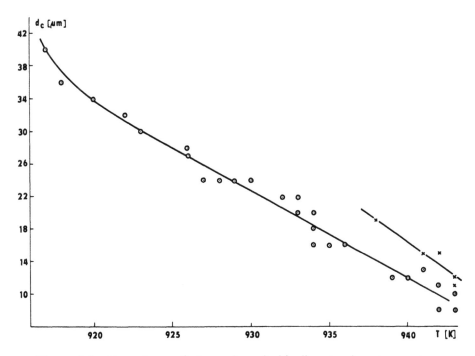

Figure 5.1 Dependence of glass column inside diameter d_c on temperature T in drawing oven for two different types of glass.

were successfully applied in gas chromatography, they are suitable for packing by a sorbent in microcolumn chromatography (Sec. 3.2.3), and they are also used in both liquid chromatography and supercritical fluid chromatography. Their preparation, that is, drawing, is based on the same principle as the drawing of glass capillaries. In the hot zone of the melting oven (Fig. 5.2) the initial fused silica tube is drawn to a capillary of the required outside and inside diameters. The raw capillary is taken out of the oven and inserted in a liquid polymer (most frequently polyimide) with which it is covered with a protective layer of suitable thickness. The process of drawing is completed in a stabilization oven, and then the fused silica capillary is cooled at a corresponding speed. The capillary is wound in a coil, which simultaneously ensures the motion of the capillary through the device.

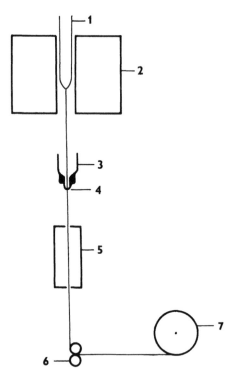

Figure 5.2 Device for drawing of fused silica capillaries: (1) initial tube; (2) graphite oven; (3) reservoir of covering polymer; (4) wetting aperture; (5) electric stabilization oven; (6) inserting silicone rollers; (7) winding wheel.

The important dependence of column inside diameter on the drawing temperature may cause the column inside diameter not to be equal at all points of the column. To verify the described method of glass capillary preparation, the value of the glass capillary column diameter was checked in several columns by breaking the column and measuring the diameter by microscope [17]. The usual method of diameter determination at the beginning and at the end of the column was considered insufficient with respect to the extremely small diameters of the capillaries.

Table 5.1 Diameters of Capillary Columns Measured by Microscope and Calculated According to Equation (5.1)

| | Column d_c (μm) | | | | | |
| | Measured by | | σ^2 | σ^2 | L | |
No.	Microscope	Calculated	(μm^4)	(%)	(m)	Mobile Phase
1	5	4.8	0.13	2.70	1.00	H_2O + 2% MeOH
2	8–9	8.0	0.17	2.72	2.20	Isoproylalcohol
3	8–9	7.3	0.38	5.20	2.20	H_2O + 5% MeOH
4	8–9	7.7	0.23	2.98	2.20	H_2O + 5% MeOH
5	8–9	8.6	0.31	3.63	1.65	Cyclohexane
6	8	8.5	0.12	1.47	2.70	Cyclohexane
7	13	11.4	0.49	4.36	1.40	MeOH
8	13	13.7	0.68	5.97	1.60	MeOH
9	13	11.3	0.32	2.83	2.06	Cyclohexane
9a	6–8	13.7	0.37	2.71	2.05	Cyclohexane
10	15	16.3	0.34	2.08	3.60	H_2O
11	17	16.9	1.24	7.30	2.05	Cyclohexane
12	17	17.1	1.73	10.10	1.96	Cyclohexane
13	34	37.6	11.77	31.31	2.00	Cyclohexane

Calculation of the capillary column diameter was carried out by a relationship derived from Poiseuille's law:

$$t_M = \frac{\varphi \eta L^2}{d_c^2 \, \Delta_P} \tag{5.1}$$

where φ is the column resistance (for capillary columns $\phi = 32$), η the mobile-phase viscosity, L the column length, d_c the column diameter, and Δ_P the pressure drop in the column.

We experimentally measured the dead elution time t_M (s), the capillary column length L (m), and the pressure at the column inlet Δ_P (MPa). We further used the values $\phi = 32$, $\eta_{H_2O}^{295.7} = 0.0942$, $\eta_{MeOH}^{295.7} = 0.0548$, $\eta_{C_6H_{14}}^{295.7} = 0.0306$, $\eta_{i-PrOH}^{295.7} = 0.195$, and $\eta_{cyclohexane}^{295.7} = 0.0710$ (values of η in Pa · s). Table 5.1 shows that the agreement between the diameter determined at the beginning of the column by microscope and that calculated by Equation (5.1) is satisfactory. Each value of d_c was calculated as an average value of measurements carried out at about 10 different mobile-phase velocities (10 different pressures at the column inlet). The standard deviation of individual measurements is on average 3.6% of d_c. Such measurements were also carried out on different parts of the measured columns and on selected columns with different mobile phases; the standard deviation ranged from 6 to 8%.

The dependence of the height equivalent of the reduced plate on the reduced velocity for the nonsorbed solute (isopropyl alcohol) on columns with diameters of 15, 14, 9, and 5 μm is given in Figure 5.3. It is evident that for columns with a diameter greater than 9 μm the curve of the dependence h on ν is shifted compared to the theoretical curve, the experimental values of h are about four times higher, and they approach the theoretical values only for the curve minimum and for lower values of ν. For a column of diameter 5 μm the results obtained do not correspond to the theory even if other measurement conditions are exactly identical. Although here the values of h also approach theoretical values for the reduced velocities, for the minimum dependence and higher linear velocities the difference between the theoretical and experimentally determined values of h increases sharply. The main reason is that the demands on detection are so high that they cannot be fulfilled in the arrangement of the apparatus elements given here [17].

The results [10,17] have shown that glass is a suitable material for the preparation of capillary columns for liquid chromatography. It can be used for the preparation of columns of sufficient length with an inside diameter of 5 μm and higher. Even at the lowest values of d_c the diameter does not change with the column length. If the extracolumnar contributions to the nonsorbed solute dispersion correspond with the requirements connected with a given column diameter, with glass columns, it is possible to obtain approximately the theoretical values of the theoretical plate height equivalent.

5.2.2   Treatment of the Column Inside Surface

Etching of the glass capillary inside surface can be done in the gas phase by hydrogen chloride, hydrogen fluoride, vapors of methyl-2-

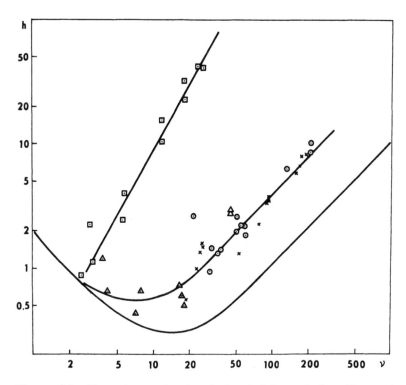

Figure 5.3 Dependence of reduced plate height equivalent H on reduced velocity v for columns with inside diameter 15 µm (circles), 14 µm (crosses), 9 µm (triangles), and 5 µm (squares).

chloro-1,1,2-trifluoroethyl ether, and the like or in the liquid phase by acid or basic agents [10,31]. Treatment in the gas phase is technically less demanding because the capillary can be packed and the surface activated with relatively low pressures of the inert gas, around 1–2 MPa. For packing the capillaries with liquids, the working pressures must be increased up to 10–15 MPa, and the probability of choking the capillaries during packing or washing is greater.

The device used for packing the capillaries by gas, vapor, or liquid has a small working volume because the capillary volume is also small (up to 10 µl) but it can withstand even higher working pressures. For packing a capillary column with a liquid the glass capillary outlet is provided in a glass pressure vessel by a short polyethylene capillary with the necessary measured volume of liquid. By increasing the pressure the liquid is forced into the column and the forcing velocity is shown by the drop in the liquid level in the polyethylene capillary. The capillary serves as a pressureless transparent container for the liquid.

The capillary column outlet is provided by another polyethylene capillary that gathers the liquid forced out by the gas.

The working conditions—solute concentration, forcing velocity, and working pressure—are dictated by the capillary inside diameter. The capillary inside surface linearly decreases with decreasing diameter; the volume of the liquid or vapor in the capillary decreases with the square of the diameter. Therefore, the decrease must be compensated for by increasing the concentration or repeating the process if the thickness of the stationary-phase film or that of the adsorption layer on the capillary wall is to be the same as in columns of larger diameter.

Fused silica capillary columns are preferred not only for their advantageous mechanical properties (elasticity and impact resistance) but also because the surface is more inert to sorption than that of glass columns. However, they also have on the inside surface sorptionally active centers that must be removed. These mostly consist of cations, such as Al, Fe, Na, and K, sometimes also Cu and B, but especially hydroxyl groups (OH) whose surface concentration corresponds to approximately one OH group on 10 nm^2. In the drawing of fused silica capillaries, nitrogen from the air can sometimes penetrate the surface and create further sorptionally active centers. Such centers participate in increasing the surface tension, which is usually around 50 mN/m for untreated fused silica capillaries.

Treatment of the capillary inside surface consists of the following gradual changes:

1. The acid groups and impurities are removed from the surface.
2. A suitable agent (most frequently trimethylchlorosilane or hexamethyldisilazane) reacts chemically with the silanol groups on the surface to which the hydrocarbon radicals are bonded

and consequently, the critical surface tension decreases to about 20–40 mN/m.

3. The wall surface is coated with a polymer liquid phase, and the necessary sorption properties of a stationary phase are obtained.

4. As a consequent of the properties of the liquid stationary phase in the liquid-liquid system described later, the stationary phase must be immobilized by cross-linking the polymer by free radicals (most frequently by azo-*t*-butane).

5. The sorption properties of the column are stabilized by washing the column with a suitable set of solvents that forms an series.

Individual procedures are described in detail in the literature [31] and if necessary to the discussion they are also presented later in this text.

5.2.3 Capillary Columns in the Liquid-Adsorbent System

One of the basic problems connected with the preparation of a reliable capillary column for liquid chromatography consists of fixation of the stationary phase on the column inside surface. The diameter of the capillary column also influences, together with the stationary phase, the number of theoretical plates necessary for the given separation, n_{req}. It is known [10] that $n_{req} = f(\beta)$; that is, the required number of theoretical plates is a function of the ratio between the two phases β:

$$n_P = 16R_s^2 \left(\frac{r}{r-1}\right)^2 \left(\frac{\beta}{K_D} + 1\right)^2 \tag{5.2}$$

where K_D is the distribution constant, r the relative retention, and R_s the resolution.

If identical chromatographic systems are used in packing microcolumns and capillary columns, the relationship $K_D = K_P = K_T$ applies (K_P is the analyte distribution constant corresponding to the sorbent and mobile-phase system in the packing column and K_T is the same constant corresponding to the system of the stationary phase on the capillary column wall and the mobile phase in the capillary column). Under such conditions the capacity ratio ($k = K_D/\beta$) for capillary columns is proportional to the ratio β_T/β_P. At a given constant resolution R_s, the ratio of the theoretical plates required for the capillary column (index T) and for the packing column (index P) [30] is

$$\frac{n_T}{n_P} = \left(\frac{k_p}{k_T}\frac{k_T + 1}{k_P + 1}\right)^2 \tag{5.3a}$$

If this condition of identity of the distribution constants in the packing and capillary column system applies, then

$$n_T = F^2 n_P \tag{5.3b}$$

where

$$F = \frac{k_p}{k_T}\frac{k_T + 1}{k_P + 1} = \frac{\beta_T}{\beta_P}\frac{k_T + 1}{k_P + 1} \tag{5.4}$$

Identity of the distribution constants in the packing microcolumn and the capillary column was verified experimentally. The experiment [10] was carried out to verify the quality of the silica gel surface formed on the glass capillary inside wall [29].

The microcolumn was packed with a pellicular material, Corasil II (Waters Assoc., Milford, MA), with a grain size from 37 to 50 μm. This material, which has a thin layer of high surface area silica gel, was selected to make the physical characteristics of the silica gel layer identical to those of the material comprising the layer to be formed on the inside surface of the glass. The column was 2 mm in inside diameter and 500 mm long. The glass capillary column was 1 m long with inside diameter 120 μm. The capillary inside surface was leached with dilute ammonia [29], which formed the silica gel layer. The mobile phase in both cases was hexane–isopropyl alcohol, 97:3.

It is evident from the results given in Table 5.2 that the distribution constants for isomers of nitroaniline are identical both on the adsorbent Corasil II and on the prepared column as shown by the relative retentions r. However, the capacity ratios for the capillary columns are substantially lower as follows from Equation (5.4). The volume of the

Table 5.2 Comparison of Capacity Ratios k and Relative
Retentions r for Packing Microcolumn and Capillary Columns

Component	k_P	r_P	k_T	r_T	F^2
o-Nitroaniline	0.44	0.23	0.0061	0.23	2,540
m-Nitroaniline	1.88	1.00	0.0265	1.00	640
p-Nitroaniline	4.36	2.32	0.0612	2.31	200

capillary column was 450 μl and the surface of the silica gel layer
0.35 m^2. These quantities correspond to the ratio $\beta_T/\beta_P = k_P/k_T = $
71.

5.2.4 Capillary Columns in the Liquid-Liquid System

One of the most important quantities in capillary liquid chromatography
is the sorption capacity related to the column length. In capillary gas
chromatography the liquid stationary phase proved to be best for several
reasons. First, it is characterized by a high variability of sorption
properties (polarity), easily adjustability of the film thickness within
acceptable limits, and consequently, also, adjustability of sorption
capacity related to the column length. Another of its advantages is the
high long-term stability of the stationary film in the column. For capillary liquid chromatography in a liquid-liquid system the stability of
the liquid film proved to be the quantity markedly limiting the applicability of the system in practical analysis.

Verification of the suitability of the liquid-liquid system for capillary liquid chromatography was carried out in both a direct-phase
(1,2,3-tris-2-cyanoethoxypropane) and a reverse chromatography system (with Apiezone L, silicone phases OV-101 and SE-30, and an
aqueous mobile phase). A sample chromatogram [24] in the reverse
liquid-liquid system with OV-101 as the stationary phase and 10^{-3} M
HClO$_4$ as the mobile phase is given in Figure 5.4. It can be seen that
under these experimental conditions good separation in a relatively
short time can be reached. The chromatogram was obtained with a
newly prepared column. The efficiency and the capacity ratios of in-

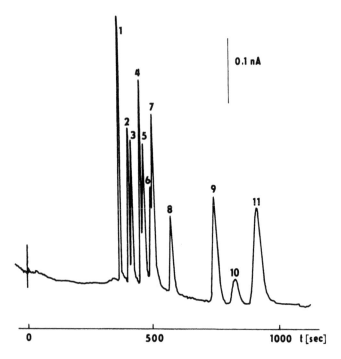

Figure 5.4 Chromatogram with capillary column of inside diameter 16 μm and 2.8 m long. Pressure 2.5 MPa, OV-101 stationary phase, mobile phase 10^{-3} M HClO$_4$ in distilled water, flow rate 1.7 nl/s. Solutes: (1) Hydrochinone (t_M); (2) methylphenol; (3) 2-methylphenol; (4) 3,4-dimethylphenol; (5) 3,5-dimethylphenol; (6) 2,3-dimethylphenol; (7) 2,4-dimethylphenol; (8) 2,6-dimethylphenol; (9) 2-methyl-4-ethylphenol; (10) 2-isopropylphenol; (11) 2,4,6-trimethylphenol.

dividual solutes decreased with time. The reasons for the instability of the efficiency and sorption capacity of capillary chromatography with a liquid stationary phase are discussed later.

5.2.4.1 *Secondary Flow of the Liquid Stationary Phase*

Capillary liquid chromatography allows us in some cases to observe and interpret effects that could not be explained with columns of larger diameters or when investigating adsorption and permeation systems with sorbents. The diameters of the capillary columns used in liquid chromatography are close to the diameters of the pores of the supports in gas chromatography and sometimes also to the diameters of the interstitial pores in liquid chromatography.

The preparation and application of capillary columns in a liquid-liquid system are connected with several effects that were observed earlier but have not been investigated closely:

1. Capacity ratios in the columns decrease with time [34], which is most probably connected with washing the liquid stationary phase out of the column.
2. Simultaneously, however, the column efficiency decreases and the height equivalent to the theoretical plate increases, although according to theory it should not. An example is given in Figure 5.5.
3. The permeability of the capillary column changes [17,34]. In some cases there is recurrent closing of the column, which can be eliminated by increased pressure [10] or sometimes by a temperature change [25]. Other times the liquid flow cannot be restored even by a pressure of several tens of MPa.
4. In some cases the column permeability and efficiency change even if the mobile phase does not flow through the column for a longer time. Such an effect was observed even in capillary columns prepared and used in gas chromatography [11]. It occasionally appears in gas chromatography, most frequently when tensides are analyzed.

The secondary flow originating on the basis of the concentration gradient of tenside in liquid-liquid or liquid-gas interfaces, called the Marangoni effect [35,36] or the Laplace-Marangoni effect, is the basis for explaining these effects [11].

The Laplace-Marangoni effect [35] is caused by the surface tension gradient at the interface [36] and its consequence is liquid flow in the interface. A liquid with a lower surface tension expands on a surface with a higher surface tension. If the surface of the liquid is curved, the convex curvature creates capillary forces that suck the liquid from

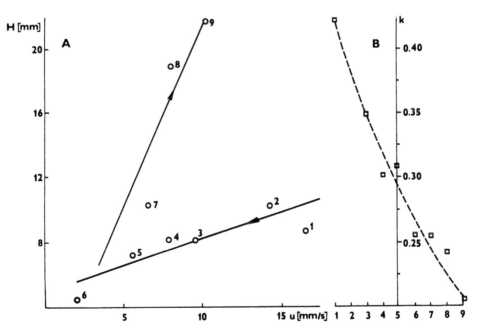

Figure 5.5 (A) Change in theoretical plate height equivalent H depending on linear velocity and sequence of experiments (numbers at points). (B) Change in capacity ratio k depending on sequence of experiments. Direction of dependence $H = f(u)$ for experiments $m_{1-6} = 0.308$ s, $m_{6-9} = 2.31$ s.

flat parts of the film to parts that are more curved (the Laplace effect). As a result of such a flow, tension that initiates a counterflow is generated at the interface (the Marangoni effect). Such flow has been described mathematically several times (e.g., Ref. 37). However, the application of mathematical procedures to individual cases is very difficult, if it is possible at all.

The Laplace-Marangoni effect can be the result [38] of the concentration gradient, the thermal gradient or the electrical charge density gradient whose consequence is the surface tension gradient on the interface. All these effects can appear in chromatography. The last, for example, has been described [39] in connection with electrokinetic effects in liquid chromatography. The Laplace-Marangoni effect may cause changes in the interfacial surface magnitude [40] and consequently also a change in the liquid film thickness [41,42].

5.2.4.2 Decrease in Capacity Ratios

Nonhomogeneity of the surface treatment or of the surface properties of the capillary column inside surface can be the cause of nonhomogeneity of the stationary-phase film thickness. The presence of surface-active substances in the chromatographic system stationary phase-mobile phase-solute leads to Laplace-Marangoni flow, increasing differences in the film thickness. Under such circumstances stationary-phase droplets may originate that are caught by adsorption forces on the surface of the column whose volume increases [43]. As a result of constant droplet increase the coating angle changes up to a critical value at which the droplet or part of it is loosened and makes a ball of diameter d_b. The inhomogeneity of the stationary-phase thickness and the tendency to create droplets were observed in glass capillary columns for gas chromatography [44]; an example of such droplets is given in Figure 5.6.

Observation of further development of the hydrodynamic system of the capillary column is based on the assumption that the droplet diameter, which depends on the surface properties of both the stationary and mobile phases, can have the values $d_f \leqslant d_b \leqslant d_c$. Assuming that $d_b \ll d_c$, stationary-phase balls can be relatively easily carried by mobile-phase flow. They can be caught on stationary-phase droplets in the column, increase their volume, and later again be released as balls to the mobile-phase flow. This creates a kind of false retention mechanism. The velocity of the decrease in the stationary-phase amount does not correspond to the velocity of the mobile-phase flow. In some cases the stationary phase is stabilized to such an extent that the droplets cannot further increase to leave the place in which they are adsorbed, and the decrease in the stationary phase from the column is considerably slowed. If under such circumstances before injection into the column the mobile is saturated by the stationary phase, the decrease in the stationary-phase volume is substantially lower; nevertheless, the largest

Figure 5.6   Scanning electron micrographs of stationary-phase droplets [44].

Table 5.3   Change in Column Efficiency

Sequence of Experiments	k_{rel}	h/γ
1	1.00	0.47
2	1.00	0.34
3	0.78	0.07
4	0.78	0.09
5	0.78	0.17
6	0.79	0.16

droplets in the stationary phase are washed out of the column. However, the surface created has fewer inhomogeneities, and consequently, the column efficiency increases to a constant value that does not change visibly over time. Experimental verification of this effect is given in Table 5.3.

If $d_b < d_c$ is assumed, then the droplet formed on the stationary-phase film surface may narrow the cross-sectional flow of the column. Under such conditions the column permeability changes, and at lower inlet pressures of the mobile phase to the column the flow may be interrupted. By increasing the inlet pressure flow can be restored with the original dependence of the flow rate on the pressure. Apparent constriction of the cross-sectional flow was shown in Reference 17. If $d_b \simeq d_c$, the droplet on the surface of the stationary-phase film is not released in the form of a moving ball but creates a column of liquid. Under certain circumstances portions of the adsorption forces and the surface tension enable penetration of the mobile phase or the mobile phase with the solute through the liquid column. This penetration is carried out in a manner often described in connection with the Maragoni effect. The moving heterogeneous liquid (or gas) penetrates through the column of the stagnant liquid, which does not change its position in the column during penetration. This differentiates penetration from bubble flow. This effect was observed visually with the penetration of nitrogen through a column of water at the end of the glass capillary column of inside diameter 100 μm.

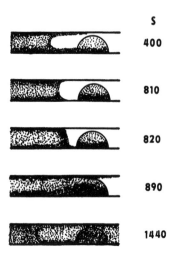

S

400

810

820

890

1440

Figure 5.7   Penetration of nitrogen through a stationary-phase column.

The dynamics of the formation of inhomogeneities of the stationary-phase film were observed photographically to depend on time [45]. We worked with a glass capillary column of inside diameter 100 μm coated with Apiezone K of layer thickness 3.2 μm. After injecting a surface-active substance (ethoxyphenol) into the column stationary-phase droplets were formed. They gradually increased and formed a stationary-phase column. The whole process was accompanied by the flow of nitrogen. Figure 5.7 shows the penetration of nitrogen through the stationary-phase column, which does not change its position in the capillary column. At the end of the experiment water was injected into the column as the mobile phase for liquid chromatography. In a relatively short time the film was destroyed and the phase was washed out of the column. The destruction process can be slowed by efficient hydrophobicization of the glass surface, which was not carried out in this case. The process of film destruction using nitrogen as the mobile phase took several hours; destruction of the film by the liquid was carried out in several tens of minutes.

In studies using a flame ionization detector [17,25] with capillary

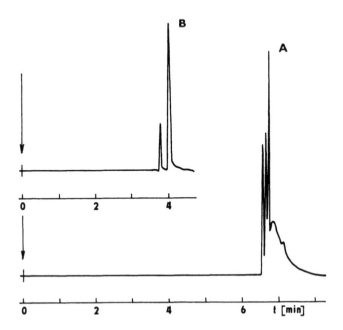

Figure 5.8 Chromatograms. (A) Elution of the solute (1,2-dibrom-2,2-dich-loroethyldimethylphosphate) after destruction of stationary-phase film. (B) Normal elution of the solute (trimethylester of phosphoric acid) from the same column under equal working conditions: flame ionization detector with alkali metal, column $L = 3.5$ m $d_c = 16$ μm, DC-550 stationary phase, mobile phase water-methanol 1:1 + 50 ppm nitrobenzene.

columns this effect could be observed on the detector response. The detector reacts quickly to the change in the flow rate. Figure 5.8A shows that the solute flow at the column outlet is interrupted at relatively regular intervals. The stationary phase in the column, DC-550 in this case, is partly dissolved by the solute and carried to the end of the capillary. Here a stationary-phase column is formed and the solute penetrates it in discrete quantities. If the liquid column is not formed, the solute is eluted from the column without the previously mentioned effect. In some cases the flow was interrupted and could not be restored even by a pressure of 40 MPa. However, if at the end of the column a piece several mm long was broken away, flow was restored in the original extent. It is assumed that in this case the stationary-phase column was so long that it did not enable penetration of the mobile phase. In other cases the interrupted flow could be restored by increasing the column temperature. In yet other cases the mobile-phase flow through the column was definitively interrupted. It is evident from these experiments that interruption of the mobile-phase flow is not caused by choking the column mechanically. The reproducibility of these experiments is relatively small because the changes in the form of the stationary-phase film can be in practice neither predicted nor controlled.

5.2.4.3 Decrease in Column Efficiency Depending on Decreases in the Solute Capacity Ratio

The incomplete continuity or inhomogeneity of the stationary-phase film thickness causes a decrease in the column efficiency. This change in efficiency is evident in Figure 5.5. A decisive change in the slope line of the can be observed for the series of experiments 1–6 and 6–9. The decrease in capacity ratios was monotonic. The changes in the shape of the film caused by the flow led to the decrease in the stationary phase and, consequently, also the decrease in the efficiency corresponding to the decrease in capacity ratios in the column. In assessing this effect we began with the known equation [7] for the reduced plate height equivalent,

$$h_t = \frac{2}{v} + \left[\frac{1 + 6k + 11k^2}{96(1 + k)^2} + \frac{k}{(1 + k)^2 d_c^2} \frac{D_m}{D_s} q d_f^2 \right] v \quad (5.5)$$

which can be expressed in simplified form by the coefficients of mass transfer in the mobile phase C_m and the stationary phase C_s and by the coefficient of the longitudinal difusion B, as follows

$$h_t = \frac{B}{v} + (C_m + C_s)v \quad (5.6)$$

In Equation (5.6) some quantities are experimentally easy to determine—v, the mobile-phase reduced velocity, k, the capacity ratio, and

d_c, the column inside diameter. Others can be found from Tables—D_m and D_s, the diffusion coefficients of the solute in the mobile and stationary phase. The product qd_f^2 (the configuration factor q and the stationary-phase film thickness d_f^2) is experimentally determined with great difficulty: it always contains a considerable amount of uncertainty with respect to the momentary state of the stationary film, which is hard to define. We assume that the product qd_f^2 is simultaneously the quantity that reflects not only the change in the stationary-phase amount in the column but also the change in the film shape. Therefore, we began by comparing the theoretically calculated value of h_t with the experimentally determined value of the reduced plate h_e. The value h_e is the sum of the columnar contribution to the height equivalent of the theoretical plate h_c and the extracolumnar contribution to the peak spreading h_{ec}:

$$\Delta h = h_e - h_t = h_c + h_{ec} - h_t \tag{5.7}$$

Equations (5.5) and (5.6) describe peak spreading on the column, and $k = 0$, $C_s = 0$, for the nonsorbed solute $h_c - h_t = 0$, and

$$\Delta h = h_{ec} \tag{5.8}$$

For $k \neq 0$,

$$\Delta h = h_e - h_t = (C_{se} - C_{st})v$$

$$= ([qd_f^2]_e - [qd_f^2]_t)v \frac{k}{(1 + k)^2} \frac{D_m}{D_s d_c^2} \tag{5.9}$$

where the indices e and t indicate the experimental and theoretical values. Whereas in all cases $[qd_f^2]_e \gg [qd_f^2]_t$ and neglecting $[qd_f^2]_t$ was not reflected in Equation (5.9) by more than about 0.5%, we can write

$$d_f q^{1/2} = \left[\Delta h \frac{(1 + k)^2}{kv} \frac{D_s d_c^2}{D_m} \right]^{1/2} \tag{5.10}$$

This experiment was carried out with glass capillary columns 1–5 whose parameters are given in Table 5.4. The columns without presaturation of the mobile phase by the stationary phase exhibited for several days a decrease in relative capacity ratio k_{re1} in the interval

Table 5.4 Parameters of Capillary Columns[a]

Column No.	Length (m)	ID (μm)	Surface Modification	Stationary Phase
1	7.5	31	TMCS[b]	Apiezon K
2	5.5	18	TMCS	Apiezon K
3	6.0	16	TMCS	Apiezon K
4[c]	6.0	16	TMCS	Apiezon K
5	5.6	15		Apiezon K
6[d]	5.1	15		DC-550
7	20.0	100	TMCS	OV-101
8	20.0	100	TMCS	OV-101

[a]Mobile phase: $H_2O + 10^{-3}$ M $HClO_4$.
[b]Trimethylchlorosilane.
[c]Saturated by stationary phase.
[d]Leached by 20% HCl, washed out with 1% HCl, and dried in H_2 at 520 K.

from 1 to 0.23 ($k_{rel} = k_n/k$, where k_n is the capacity ratio of the component in the test n). If the decrease was not as important, the column was spontaneously closed and the mobile-phase flow interrupted before the experiment ended. These experiments were carried out without disassembling the entire apparatus even if only one column was used. Therefore, the extracolumnar contributions to the peak spreading in one test can be considered constant and all the changes in the system efficiency can be explained only by changes in the shape and amount of the stationary phase in the column. An example of experimental verification of Equation (5.10) is given in Figure 5.9.

With the decreasing capacity ratio k_{rel}, the stationary-phase volume V_s also decreases. If the distribution constant K_D is assumed to remain unchanged, the thickness of the stationary-phase film d_f corresponding to the homogeneous distribution of liquid on the surface of the column inside wall also decreases as a result of the decrease in the total amount of the stationary phase on the column surface:

$$V_{s,rel} = k_{rel} \frac{V_m}{K_D} = \pi d_c d_{f,rel} L \qquad (5.11)$$

Despite this, the experimentally obtained value $q d_f^2$ increases. In agree-

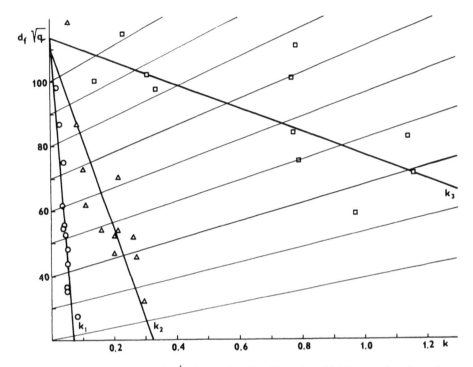

Figure 5.9 Quantity $d_f\sqrt{q}$ determined by Equation (5.10) as a function of the change in capacity ratio k for three different solutes.

ment with this assumption [Eq. (5.8)], $d_fq^{1/2}$ is constant for $k = 0$ irrespective of the initial capacity ratio of the solute. The experimental dispersion variation in the values is $\pm 1.9\%$ and the mean value for $d_fq^{1/2}/d_c = 3.16$ for columns with inside diameter 15–31 μm (1–6), which indicates that the extracolumnar contributions to peak spreading did not differ very much.

These experiments further show that the functional dependence of the reduced plate height equivalent on the capacity ratio differs from that assumed from Equation (5.5), from which the dependence (5.10) is also derived. Figure 5.9 shows that the slope of the line connecting the values $d_fq^{1/2}$ for k_1, k_2, and k_3 increases with the decreasing relative capacity ratio of the solutes. These slopes are given in Table 5.5

Table 5.5 Dependence $d_f\sqrt{q} = f(k)$ According to Equation (5.10)

k_{rel}	Slope	Correlation Coefficient
1.00	40.16	1.00
0.98	42.42	1.00
0.93	85.52	0.93
0.78	56.95	0.99
0.77	75.20	0.97
0.42	129.78	0.97
0.42	107.89	0.96
0.34	156.00	0.97

together with correlation coefficients that depend on the relative capacity ratio. The results of this experiment prove that the capacity ratios of the solutes decrease with time and simultaneously the stationary-phase film inhomogeneity increases. The inhomogeneity increase is more significant for solutes with higher capacity ratios.

5.2.4.4 Importance of Stationary-Phase Secondary Flow

It has been shown that the amount of the liquid stationary phase and its distribution in glass capillary columns in the presence of surface-active substances changes with time. The Laplace-Maragoni flow generates inhomogeneities in the film thickness, for example, stationary-phase droplets, which cause a decrease in the column efficiency and in the stationary-phase content in the column. Consequently, the inhomogeneity of the stationary-phase distribution increases with the increasing capacity ratio of the solute in the column.

Application of tenside as a solute on the column also causes a decrease in the capillary column efficiency in gas chromatography. This process takes place even without carrier gas flow.

Improvement in the ratio between the volume of the stationary and mobile phases can be made by decreasing the column inside diameter. In a liquid-liquid system the limit column diameter d_c most probably equals 5 μm. Columns with lower diameters exhibit, with respect to the previously mentioned problems, defects in mobile-phase flow and a rapid decrease in efficiency. In using adsorption systems the generation of two heterogeneous liquid phases on the column, although temporary, must be prevented. Saturation of the mobile phase by the stationary phase can stabilize the capacity ratios on the column; even under such circumstances column efficiency nevertheless decreases.

The pore dimensions of stationary-phase supports for gas chromatography and the interstitial pore dimensions in liquid chromatography are close to the inside diameters of the columns studied. Therefore, the influence of Laplace-Marangoni flow in the stationary or stagnant phase on the stability of chromatographic column packing cannot be excluded.

The results of the study of Laplace-Marangoni flow in glass capillary columns show that the liquid-liquid system is quite unsuitable for capillary chromatography. A partial solution can be seen in the adsorption system with a stationary phase bound chemically to the surface. A disadvantage of this solution is the relatively small sorption capacity of the surface to a unit of column length. A further possible solution is the application of cross-linked polymers as the stationary phase. However, in neither case can the effects connected with Laplace-Marangoni flow be excluded.

5.2.5 Capillary Columns with an Immobilized Stationary Phase

With respect to the properties of the surface tension gradient, systems based on the coexistence of two liquid phases, one of which is considered stationary because of its higher sorption affinity to the capillary wall, are regarded as entirely unsuitable for capillary chromatography. Coating with high-molecular-weight polymers also does not prove successful. Liquids, on the contrary, enable easy regulation of selectivity and absorption capacity, which are highly desired properties for the stationary phase.

Chemical bonding of the liquid molecules on the surface of the capillary column wall can in principle be considered immobilization of the stationary phase. It has been shown that chemical bonding is possible with both low-molecular-weight monomers (e.g., octadecylsilanes, which prove successful on silica gel sorbents in reversed-phase packing columns for a long time) and silicone polymers. A deficiency of chemically bonded monomers is their low sorption capacity with respect to the low values of the geometric surface of the capillary wall. Viscous polymers can be injected into a capillary of small diameter only with difficulty: their viscosity and structure decrease the diffusion coefficients by several orders of magnitude. A diffusion coefficient of about 10^{-8} cm^2/s increases the mass transfer coefficient in the stationary phase and makes application of the capillary column impossible at the mobile-phase velocities necessary to ensure that the time of analysis is short enough. Chemical bonding of the stationary phase to the column surface can be obtained by treating the capillary surface by leaching. As has been discussed, leaching of the columns of diameter <15 μm is connected with the danger of choking the column as a result of gel or salt formation inside the column.

Immobilized stationary phases in capillary chromatography, based mostly on silicone, are those that are bonded to the capillary wall by chemical reaction and cross-linked inside the column. The extent of cross-linking of the immobilized phases usually ranges from 0.1 to 1%, which ensures relatively high diffusion coefficients (up to 10^{-6} cm^2/s) and good dynamic properties of the stationary phase. The thickness of the immobilized stationary phase can be regulated from tens of nm to several μm. This simultaneously enables us to regulate the ratio of the phase volume in the column. V_m/V_s decreases below 10, which again positively influences the dynamic qualities of the capillary column.

Cross-linking of immobilized phases is at present carried out with four basic types of initiation agents [46]: azoalkanes, alkylperoxides, gamma radiation, and ozone. Azo-t-butane [(CH$_3$)$_3$C—N=N—C(CH$_3$)$_3$] is particularly convenient for preparation of immobilized phases. It can be introduced into the column in the form of vapor. Its decomposition temperature is 180–220°C. Another frequently used cross-linking initiator is dicumylperoxide, C$_6$H$_5$(CH$_3$)$_2$C—O—O—

C(CH$_3$)$_2$C$_6$H$_5$ [47]. Its decomposition temperature is 150–180°C. Because it is a strong oxidation agent, the cross-linking effect is sometimes accompanied by changes in the chemical character of the column wall surface, which is sometimes reflected in undesirable sorption activity of the resulting surface.

With an immobilized stationary-phase thickness of 0.25 μm and above, the influence of the column wall is considered negligible. The relative retentions remain constant, and with increasing stationary-phase thickness, that is, with increasing stationary-phase volume, the capacity ratio also increases proportionally.

The sorption properties of immobilized stationary phases can be adjusted by adding modifiers to the mobile phase. In mobile-phase saturation with paraffinic hydrocarbons (heptane), the capacity ratios increased as much as 10 times. Under such circumstances the methylsilicone phase swells and its volume increases nearly 3 times. The stationary phase used for these solutes is usually hydrocarbon in the form of cross-linked polymer. Its swelling enables further efficient regulation of the stationary-phase volume as well as control of the phase volume ratio.

At present immobilized stationary phases seem to represent the most important technique of preparation of capillary columns for liquid chromatography. Their long-term stability and good resistance to a number of solvents and, therefore, their applicability to gradient techniques, together with their relatively easy preparation predisposes them for wide application.

5.2.6 Other Types of Capillary Columns

In the capillary column the chemically bonded phenylsilyle groups or aliphatic chains can be sulfonated. This enables ion exchangers with an —SO$_3$H active group. These exchangers on the chemically bonded aromatic structure are analogous to exchangers created on styrene-divinyl copolymer that are currently used in ion-exchange chromatography. Although the exchange capacity of the granulated exchangers ranges from 10^{-3} to 10^{-6} mol/g, in glass capillary columns the capacity usually equals 10^{-8} mol/m [23]. This increase in the exchange capacity is connected to the increase in the surface magnitude and, therefore, with the technique of glass leaching before chemical modification.

Great variability in the selectivity of the columns can be obtained with dynamically modified stationary phases. This technique is used both in adsorption systems with silica gel on the column wall and in reversed base systems. The surface is modified mostly by ionogenic surface-active substances that when the silica gel layer is applied hydrophobicize the surface; the reversed-phase systems, in contrast, reduce the surface hydrophobicity. In some cases they may also act as ion exchangers. The concentration dependence on the retention of the given solutes on the modifier concentration, which is similar to the dependence obtained on packed columns (see Fig. 4.3), enables us to change the capillary column selectivity over a sufficiently wide range.

A common feature of all types of capillary columns is their extremely low sorption capacity. Consequently, only very small amounts of solutes can be applied to columns, which places extreme demands not only on the injection technique but also on the detectors.

5.3 DETECTORS

It was theoretically derived [7,9] that at constant chromatographic parameters, such as n, D_m, and ΔP for packing and capillary columns, the capillary column diameter must be 5.5 times lower than the sorbent particle diameter to carry out the analysis in an equal amount of time. Because it is necessary to maintain the relation between the column variance $\sigma_{V,c}^2$ and the extracolumnar variance, in this case the detection variance $\sigma_{V,\text{det}}^2$, it is evident that the demands on the magnitude of the detection cell volume are higher in capillary chromatography than in microcolumn chromatography.

Figure 4.3 practically demonstrates the importance of the detector volume [17]. For columns of inside diameter from 15 to 9 μm the detectors used here are satisfactory [24,25]; for a column of inside diameter 5 μm the character of the dependence $h \sim v$ is quite different.

Table 5.6 Required Volume of Detection Cell
Depending on Column Diameter[a]

d_c (10^{-6} m)	$\frac{1}{2}\,\sigma_{V,c} = V_{det}$ (m^3)	V_{det}
1	3.9×10^{-16}	0.39 pl
5	2.2×10^{-14}	22.00 pl
10	1.2×10^{-13}	0.12 nl
50	6.9×10^{-12}	6.90 nl
100	3.9×10^{-11}	39.00 nl

[a]According to Equation (5.13), selected values $L = 1$ m, and $h = 1$.

For detector cell volume used the known relationship (e.g., Refs. 7 and 24) is

$$V_{det} = \frac{1}{2}\,\sigma_{Vc} = \frac{d_c^2\,L}{8n^{1/2}} \tag{5.12}$$

If in Equation (5.12) the number of theoretical plates n is substituted by the value h, the following relationship for the standard deviation $\sigma_{V^{\circ}C}$ is obtained:

$$\sigma_{V^{\circ}C} = \frac{\pi}{4}\,d_c^{5/2}h^{1/2}L^{1/2} \tag{5.13}$$

Suppose that the value h in the chromatographic system depends only on the column properties and is not affected by the magnitude of the detection cell. For capillaries of inside diameters corresponding to good conditions of analysis (sufficiently short time of analysis and sufficiently high peak capacity [9,48]), the requirements for the detection cell volume cannot be fulfilled under the present conditions. The selected examples for $L = 1$ m and $h = 1$ are given in Table 5.6. The situation is technically unsolvable especially for spectrophotometric detectors with capillary cuvettes oriented against the light in the direction of the longitudinal axis. For example, for a column of diameter $d_c = 5$ μm, a cell whose channel would have the diameter 8×10^{-2} μm for an optical length of only 0.4 mm would have to be designed. From Equation (4.13) is also evident that the requirements for the detection cell volume cannot be markedly reduced by making the column longer. If the column is 10 times longer, the maximal cell volume increases only 3.16 times. The allowable cell volume increases with increasing capacity ratio of the solute $1 + k$ times.

These data are probably why spectrometric and fluorimetric detectors work mostly with cells of volume 50–100 nl [2,49–52]. A cell with a volume of 6 nl can be designed [53] only at the expense of substantial shortening of the optical path. In this case the fused silica capillary is examined against the light perpendicularly to its longitudinal axis.

Some electrochemical principles of detection allow the design of very small detection cells. Application of suitably thin wires [7,24] as electrodes enables us to assemble amperometric detectors of the wall-jet type with a cell volume of several nl [54–57] and even with a cell volume of less than 1 nl [24]. In the oxidation mode with the platinum electrode (Fig. 5.10A), the amperometric detector obtains good efficiency. The minimum detectable concentration ranges [58] around 10^{-6} g/L (Table 5.7). For nonsorbed hydroquinone the minimum detectable amount of 5×10^{-16} mol was determined with a glass capillary column of inside diameter 14 μm and length 1.2 m. Such a value is close to that obtainable by potentiometric detection (Fig. 5.10B) with capillary columns [27], which is 6×10^{-15} mol.

A potentiometric detector with an ion-selective electrode [27,28] exhibits a geometric volume of 0.5 fl (10^{-16} L), assuming a spherical diffusion layer. The same detector exhibits an effective volume from 8 to 20 fl assuming a diffusion layer of thickness equalling the column diameter. In practical measurements values from 53 to 520 fl were obtained on evaluation of the response of the nonsorbed solute. It is evident that the character of the flow in the detection cell, as well as the detection mechanism, play an important role, and therefore, the time constant of the detector response [17,24,27,59] is used in some cases as a more objective criterion.

Table 5.7 Minimum Analyzable Amounts and Minimum
Detectable Concentrations for the Solutes[a]

		Minimum Analyzable Amounts (pg)	Minimum Detectable Concentrations (μg/L)
Sulfonamides	Sulfanilic acid	5	1.3
	Sulfanilamide	6	1.5
	Sulfacetamide	20	3
	Sulfathiazole	10	6
	Sulfamethoxydiazine	60	5
Phenothiazines	Levopromazine	40	3.5
Thioxanthene	Chloroprotixen	500	44
	Chloropromazine	40	2.5
	Thioridazine	60	2.7
	Prochloroperazine	100	3.5
Tetracylclines	Rolitetracycline	500	47
	Tetracycline	1,000	49
Chlorinated	4-Chlorophenol	20	3
phenols	2,4-Dichlorophenol	30	4
	2,4,6-Trichlorophenol	50	5
	2,3,4,6-Tetrachlorophenol	200	—
	Pentachlorophenol	200	11
Condensed	Anthracene	30	3.3
aromatic	Pyrene	100	7
hydrocarbons	Perylene	40	1.7
	1,2-Benzpyrene	40	1.5
	20-Methylcholanthrene	90	2.5
Parabenes	4-Hydroxybenzoic acid	300	83
	Methyl-4-hydroxybenzoate	300	68
	1-Propyl-4-hydroxybenzoate	900	140
Vitamins	Vitamin A acetate	200	23
	Vitamin D_2	800	50
	Vitamine E	12,700	409
	Folic acid	200	20

(continued)

Table 5.7 Continued

		Minimum Analyzable Amounts (pg)	Minimum Detectable Concentrations (μg/L)
Flavonoids	Quercetin	30	2.3
Azo dyes	4-Aminoazobenzene	20	4
	2-Aminoazotoluene	30	5
	N,N-Dimethyl-4-amino-azobenzene	20	3.3
Aromatic amines	Benzidine	3	0.9
	1-Naphthylamine	6	1.0
	2-Naphthylamine	5	1.2
	Carbazole	9	1.4
	Diphenylamine	6	0.8

[a]Assumed value of noise 50 pA.

The importance of the detector time constant in connection with capillary liquid chromatography stands out if reduced values [60] of basic chromatographic quantities are used. Using the relationship for the reduced column length,

$$\lambda = \frac{L}{d_c} \qquad (5.14)$$

Equation (5.13) can be expressed using the reduced values of the theoretical plate height equivalent h, the velocity v, and the length λ as

$$\sigma_{v^\circ C} = \frac{\pi}{8} d_c^3 h^{1/2} \lambda^{1/2} \qquad (5.15)$$

If the reduced detector volume [27] W_{det} is introduced, then

$$W_{det} = \frac{V_{det}}{d_c^3} = \frac{\pi}{8} h^{1/2} \lambda^{1/2} \qquad (5.16)$$

Transferring volume to time as is usual in chromatography, $V = tF$, the relationship for the detector time constant is obtained:

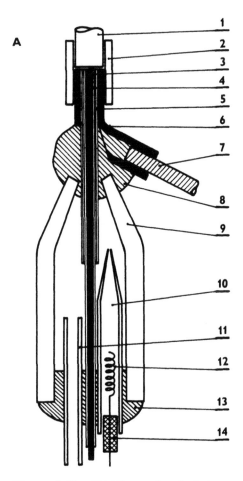

Figure 5.10 (A) Electrochemical detectors for capillary columns. Apero-metric detector according to Reference 24: (1) capillary column; (2) PTFE tube; (3) Pt wire (diameter 0.1 mm); (4 and 5) glass capillaries; (6) stainless steel capillary; (7) connection of auxiliary electrode; (8 and 13) epoxy resin; (9) glass vessel for reference electrode; (10) reference electrode; (11) outlet capillary; (12) silver wire; (14) rubber stopper.

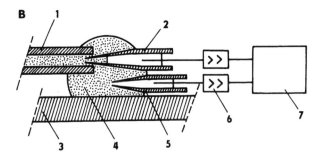

Figure 5.10 (B) Potentiometric detector according to Reference 27: (1) capillary column; (2) ion-selective microelectrode; (3) glass plate; (4) effluent droplet; (5) comparison electrode; (6) amplifier; (7) recorder.

$$t_{det} = \frac{h^{1/2}\lambda^{1/2}}{2v} \tag{5.17}$$

The detector time constant expressed with the reduced quantities shows a direct dependence of the required detector property on the parameters of the chromatographic system. This does not exclude the possibility of using detection cells of higher volumes if a sufficiently rapid detector response to the given solute can be ensured.

It has been shown [10] that capillary columns of inside diameter from 60 to 80 μm can be connected to a spectrophotometric detector with a cell of volume 8 μl. The spectrophotometric detector (LCD 254, Laboratory Instruments, Prague) was connected to the capillary column outlet through a flow splitter. The flow splitter was made from a glass T-shaped piece of suitable dimensions from inlet and outlet capillaries. The splitter was filled with the washing liquid, which washed the capillary column outlet and introduced the solute to the detector at a velocity much higher than that of the mobile-phase flow through the column. This technical solution of detection in capillary liquid chromatography is connected to the increase in the minimum detectable solute concentration at the column outlet; however, it allows us to use detectors with a detection cell volume greatly exceeding the volume admissible according to Equation (4.12). The described test [10] presented detection of the solutes leaving the column in volumes of less than 1 μl without substantial peak distortion. The technique helped to obtain 1,250,000 theoretical plates for the nonsorbed solute on a column of inside diameter 60 μm and length 21 m. The column output was approximately 50 theoretical plates per second.

In some cases a sufficiently low detection time constant is also exhibited by transport detectors. These are detection systems in which the solute is transferred from the chromatographic column outlet to the detecting element by medium other than the mobile phase. It is evident that with this type of detector the method just described with the washing liquid can be also included if its composition differs from that of the mobile phase. However, in most types of transport detectors application of the transfer medium is not connected to a decrease in sensitivity, that is, with an increase in the minimum detectable concentration or minimum detectable mass flow of the solute through the detector.

Detectors with mechanical solute transfer to the detecting element, that is, wire or chain detectors [61,62], which use a flame ionization detector or a flame ionization detector with alkali metal as the detecting element, also proved successful. In some cases the detecting element is represented by mass spectrometer [63].

The other possibility of solute transport is aerosol generation at the column outlet. The solute is transported this way either to a spectrophotometer [64–66] or to a flame ionization detector [67].

For capillary liquid chromatography pneumatic transport of the solute to the flame ionization detector [17,25,26,68,71] can also be used. The solute, after evaporation and pyrolysis, sometimes also transfer to an aerosol, is transported by a mixture of nitrogen and hydrogen to the space of the ionizing flame and detection electrodes. Solute transport is carried out at a substantially higher velocity than that of the liquid mobile phase in the column. Consequently, the detector effective volume is lower and the time constant shorter than the corresponding geometric dimensions of the detector.

A special torch was designed that was used in the flame ionization detector and in the flame ionization detector with alkali metal (Fig. 5.11). This consists of a fused silica tube to which the end of the capillary column is introduced. The capillary column outlet is washed

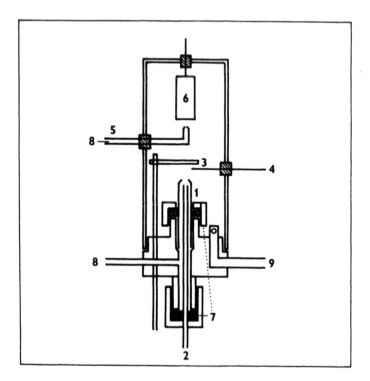

Figure 5.11 Flame ionization detector with alkali metal: (1) quartz burner of the detector; (2) capillary column; (3) supply of alkali metal; (4) auxiliary electrode; (5) metal burner of the detector; (6) cylindrical sensing electrode; (7) silicone rubber seal; (8) inlet of hydrogen and nitrogen mixture; (9) air supply.

with the hydrogen-nitrogen mixture (e.g., H_2/N_2 50:50, vol/vol). The temperature of the capillary column end can be regulated from the laboratory temperature to a temperature of about 1,000 K according to the type of torch and the distance of the capillary column end from the torch flame. Under suitably selected conditions the velocity of the solute transfer to the detector remains unchanged for solutes with a range of boiling points from 350 to 600 K.

The flame ionization detector was used in the usual design with electrometric amplifiers. The same electrometric amplifiers can be also used for work with the flame ionization detector with alkali metal in a dual arrangement [72,73].

The flame ionization detector can be also used with mobile phases to which it gives a response. If the detector response to the solute is also considered standard for water as the liquid phase, $R_i^{H_2O}$, then the detector response [74] R_i for the mobile phase containing a detectable substance b as its component (e.g., methanol) is given by the relationship

$$R_i = R_i^{H_2O} \left(1 - \sum C_b^{ef} \frac{y_b}{\sum C_i^{ef}} \right) \qquad (5.18)$$

where $\sum C^{ef}$ is the number of effective carbons of component b or of solute i and y_b is the molar fraction of component b in water. The detector response decreases with the increasing value of the fraction on the right side of Equation (4.18). As in gas chromatography, it decreases with increasing tension of the stationary phase passing from the column to the detector [71]. The detector may also provide a negative response if $\sum C_i^{ef} \ll \sum C_b^{ef} y_b$. The ionization efficiency of the flame ionization detector for isopropyl alcohol in the mobile phase consisting of water and 5% methanol was 0.02 C/mol. Similar values of ionization efficiency were also obtained for other solutes.

The flame ionization detector with alkali metal extends the possibilities of application of this detection principle for capillary liquid chromatography [72, 73]. For example, the hydrocarbon mobile phase can be used for halogen substances and substances containing phosphorus. The sufficient selectivity of the detector response ensures convenient sensitivity of detection for these substances.

Despite the advance obtained in investigation of electrochemistry and transport and also in some cases of optical detection principles, detectors remain a limiting element in the further development of capillary liquid chromatography. A number of detectors described in connection with microcolumns (Sec. 3.2.5) were used. Attention is given mainly to optical detectors using lasers.

REFERENCES

1. Hibi K., Ishii D., Fujishima I., Takeuchi T., Nakanishi T.: J. High Resol. Chromatogr. Chromatogr. Commun 1, 21 (1978).
2. Tsuda T., Hibi K., Nakanishi T., Takeuchi T., Ishii D.: J. Chromatogr. 158, 227 (1978).
3. Tsuda T., Novotný M.: Anal. Chem. 50, 632 (1978).
4. Tijsen R.: Separ. Sci. Technol. 13, 681 (1978).
5. Dewaele C., Verzele M.: J. High Resol. Chromatogr. Chromatogr. Commun. 1, 174 (1978).
6. Krejčí M., Šlais K., Tesařík K.: J. Chromatogr. 149, 654 (1978).
7. Knox, J. H., Gilbert M. T.: J. Chromatogr. 186, 405 (1979).
8. Knox J. H.: J. Chromatogr. Sci. 18, 453 (1980).
9. Yang, F. J.: J. Chromatogr. Sci. 20, 241 (1982).
10. Krejčí M., Tesařík K., Pajurek J.: J. Chromatogr. 191, 17 (1980).
11. Krejčí M., Tesařík K.: J. Chromatogr. 282, 351 (1983).
12. Hofmann K., Halász I.: J. Chromatogr. 173, 211 (1979).
13. Hofmann, K., Halász I.: J. Chromatogr. 199, 3 (1980).
14. Snyder R. L., Dolan J. W.: J. Chromatogr. 185, 43 (1979).
15. Dolan J. W., Snyder R. L.: J. Chromatogr. 185, 57 (1979).
16. Meyer, R. F., Champlin P. B., Hartwick R. A.: J. Chromatogr. Sci. 21, 433 (1983).
17. Krejčí M., Tesařík K., Rusek M., Pajurek J.: J. Chromatogr. 218, 167 (1981).
18. Golay M., in: Gas Chromatography 1958 (D. H. Desty, ed.), Butterworths, London, 1958, p. 36.
19. Ishii D., Takeuchi T.: J. Chromatogr. Sci. 18, 462 (1980).
20. Tijssen R., Bleumer J. P. A., Smit A. L. C., van Kreveld M. E.: J. Chromatogr. 218, 137 (1981).
21. Tsuda T., Tsuboi K., Nakagawa G.: J. Chromatogr. 214, 283 (1981).
22. Jorgenson J. W., Guthire E. J.: J. Chromatogr. 255, 335 (1983).
23. Ishii D., Takeuchi T.: J. Chromatogr. Sci 22, 400 (1984).
24. Šlais K., Krejčí M.: J. Chromatogr. 235, 21 (1982).
25. Krejčí M., Rusek M., Houdková J.: Collect. Czech. Chem. Commun. 48, 2342 (1983).
26. Houdková J.: Diploma Thesis, Faculty of Natural Sciences, J. E. Purkyně University, Brno, 1982.
27. Manz A., Fröbe Z., Simon W., in: Microcolumn Separations, J. Chro-

matogr. Library, Vol. 30 (M. V. Novotny, D. Ishii, eds.), Elsevier, Amsterdam, 1985, p. 297.

28. Manz A., Simon W.: J. Chromatogr. Sci. 21, 326 (1983).
29. Tesařík K.: J. Chromatogr. 191, 25 (1980).
30. Tesařík K., Kaláb P.: Chem. Listy 69, 1078 (1975).
31. Tesařík K., Komárek K.: Kapilární kolony v plynové chromatografii (Capillary Columns in Gas Chromatography), SNTL, Prague, 1984.
32. Desty D. H., Haresnape J. N., Whyman B. H. F.: Anal. Chem. 32, 302 (1960).
33. Tesřík K., Šlais K.: Technika přípravy skleněných kapilárních kolon s vnitřnín průměrem 2–30 μm pro kapalinovou chromatografii (Treatment of Glass Capillary Columns with I. D. from 2 to 30 μm for Liquid Chromatography), VZ 68, Research Report for Laboratory Instruments Prague, Institute of Analytical Chemistry, Czechoslovak Academy of Sciences, Brno, 1980.
34. Takeuchi T., Ishii D.: J. Chromatogr. 240, 31 (1982).
35. Scriven L E., Sterling C. V.: Nature 187, 186 (1960).
36. Ross S., in Encyclopedia of Chemistry (G. L. Clark, G. G. Hawley, eds.), Reinhold, New York, 1966, p. 446.
37. Scriven L. E.: Chem. Eng. Sci. 12, 98 (1960).
38. Sawistowski H.: Chem. Ing. Technol. 45, 1093 (1973).
39. Šlais K., Krejčí M.: J. Chromatogr. 148, 99 (1978).
40. Sawistowski H.: Chem. Ing. Technol. 45, 1115 (1973).
41. Dukhin S. S., in: The Modern Theory of Capillarity (F. C. Goodrich, A. I. Rusanov, eds.), Akad. emischl. Verlaggesselschft, Berlin, 1981, p. 83.
42. Ewans L. F., Ewers W. E.: Ind. Eng. Chem. 46, 2420 (1954).
43. Birkeman J. J.: Physical Surfaces, Academic Press, New York, 1970, p. 239.
44. Bauer P., Tesařík K., Pospíšil J., Komárek K.: Chem. Listy, 79, 756 (1985).
45. Krejčí M., Tesařík, Březina V.: Liquid Stationary Phase Distribution in Capillary Chromatography, Workshop on Microcolumn Liquid Chromatography, Free University of Amsterdam, 1984.
46. Borke V., Hubáček J., Řeháková V.: Chem. Listy 79, 364 (1985).
47. Grob K.: Making and Manipulating Capillary Columns for Gas Chromatography, Hütig Verlag, Heidelberg, 1986.
48. Giddings J. C.: Anal. Chem. 39, 1027 (1967).

49. Ishii D., Asai K., Hibi K., Jonokuchi T., Nagaya M.: J. Chromatogr. 144, 157 (1977).
50. Hershberger L. W., Callis J. B., Christian G. D.: Anal. Chem. 51, 1444 (1979).
51. Hirata Y., Novotný M.: J. Chromatogr. 186, 521 (1979).
52. Hirata Y., Lin P. T., Novotný M., Wightman R. M.: J. Chromatogr. 181, 287 (1980).
53. Yang, F. J.: J. High Resol. Chromatogr. Chromatogr. Commun. 4, 83 (1981).
54. Šlais K., Kouřilová D.: Chromatographia 16, 265 (1983).
55. Šlais K., Krejčí M., Pavlíček M.: PV 4635-82.
56. Štulík K., Pacáková V.: Chem. Listy 73, 795 (1979).
57. Šlais K.: J. Chromatogr. Sci. 24, 321 (1986).
58. Krejčí M., Šlais K., Kouřilová D., Vespalcová M.: J. Pharm. Biomed. Anal. 2, 197 (1984).
59. Kucera P.: J. Chromatogr. 198, 93 (1980).
60. Knox, J. H.: Liquid Chromatography Techniques in Chemistry, Vol. 11, John Wiley, New York, 1976.
61. Scott R. P. W.: Liquid Chromatography Detectors, Elsevier, New York, 1977.
62. Šlais K., Krejčí M.: J. Chromatogr. 91, 181 (1974).
63. McFadden W. H., Schwartz J. L.: J. Chromatogr. 122, 386 (1976).
64. Freed D. J.: Anal. Chem. 47, 186 (1975).
65. Gast C. H., Kraak J. H., Poppe H., Maessen F. J. M. J.: J. Chromatogr. 185, 549 (1979).
66. Julin B. C., Vandenborn H. W., Kirkland J. J.: J. Chromatogr. 112, 443 (1975).
67. Krejčí M., Tesařík K.: AO 167052-1975, U.S. Patent 4,014,793.
68. Ettre L. S.: Open Tubular Columns in Gas Chromatography, Plenum Press, New York, 1965.
69. Folestad S., Larson M.: J. Chromatogr. 394, 455 (1987).
70. McGuffin V. L., Novotny M.: Anal. Ehcm. 53, 946 (1981).
71. McGuffin V. L., Novotny M.: Anal. Chem. 55, 2296 (1983).
72. Karmen A.: Anal. Chem. 36, 1416 (1964).
73. Chundela B., Krejčí M., Rusek M.: Dtsch. Z. Gericht. Med. 62, 154 (1968).
74. Krejčí M., Dressler M.: Chromatogr. Rev. 13, 1 (1970).
75. Novák J., Gelbičová-Rüžičková J., Wičar S., Janák J.: Anal. Chem. 43, 1996 (1971).

6
Examples of Analysis

It is difficult to select examples of the use of microcolumn liquid chromatography in analytic practice. First, all the analytic procedures described for high-performance liquid chromatography (HPLC) can also be used for microcolumn liquid chromatography. This means that there is a large body of information in the literature that cannot be described here completely. The situation is further complicated by the current application of microcolumns in many analytic laboratories. The authors working with such apparatuses often do not mention such terms as "microbore," "microcolumn," and "capillary column" in the titles of their works. In many cases the term "high-performance liquid chromatography" has become synonymous with "microcolumn liquid chromatography" just as the term "trace analysis" has become synonymous with "analysis."

In the beginning of the development of microcolumn and capillary chromatographic techniques, the number of applications was limited by the lack of commercially available apparatus. This situation changed during the 1980s, and a great majority of the applications described here were carried out on commercial equipment. The only exceptions are detectors, mainly those based on laser techniques, that are still most often designed in the laboratories of individual authors.

Perhaps the most important practical applications of microcolumn liquid chromatography are those in which the advantages are demonstrated by theory. These are mainly applications that make use of the possibility of working with substantially lower amounts of analyte sample than used for conventional analytic columns and those that use a higher analyte concentration at the microcolumn outlet. Both types are reflected in the successful application of microcolumn liquid chromatography for trace analysis. We can also make use of new possibilities of combinations of liquid chromatography with spectral methods, as well as the decreased consumption of sometimes very expensive and rare stationary phases, mainly chiral stationary phases. The advantage of very low consumption of expensive mobile phases has only rarely been stressed, most probably because combinations of inexpensive chemicals can influence the phase equilibrium of the analyte in the mobile phase.

6.1 BIOLOGICAL SAMPLES

The advantage of working with a low sample amount while maintaining the detection limit is evident in trace analysis especially if higher sample amounts can be obtained only with difficulty, sometimes with unacceptable risk, or if the analyzed object does not provide sufficiently high sample volumes. An example of the first case is the examination of infants: it is of great advantage to draw only 500 µl blood for analysis instead of the 5–10 ml of sample required for conventional chromatographic columns. In some small experimental animals the blood volumes necessary for analysis with conventional apparatuses do not even exist.

Analysis of an individual cell has been for a long time an exciting task for analysts, mainly for chromatographists. This task becomes more urgent with increasing knowledge of the heterogeneity of tissues. By application of open tubular columns in liquid chromatography, cells from tissues studied in neurobiology have been analyzed [1]. The basic advantage of open tubular capillary columns for this kind of analysis is the possibility of using a typical injection volume of 5 nl. An entire column of ID (inner diameter) 15 μm and 2 m long had a total volume of only 390 nl. With such a column efficiencies comparable with those of packing columns or even greater can be obtained. To take advantage of the properties the column, a sufficiently sensitive and selective detector had to be selected. Hence a voltammetric detector with a detection limit of 0.1 fmol for hydroquinone was used. The same detector working in the amperometric mode exhibited a detection limit of 1 amol. This detector was provided with a carbon fiber microelectrode [2–5] of diameter 9 μm and 1 mm long. The potential of the electrode was ramped from 0.0 to 1.3 V versus an Ag/AgCl electrode. The voltammograms were repeated at 1.3 s intervals and performed at a velocity of 1 V/s. Retention times in combination with voltammograms can be used for identification of analytes.

For separation of tyrosine (Tyr), tryptophan (Trp), dopamine (DA), serotonin (5HT), 3,4-dihydroxyphenylalanine (DOPA), 3,4-dihydroxyphenylacetic acid (DOPAC), 5-hydroxytryptophan (5-HTP), and 5-hydroxyindolacetic acid (HIAA), a column with a reversed stationary phase was selected. The glass capillary column of inside diameter 15–19 μm and 220–250 cm long had dimethyloctadecylsilane chemically bonded on its inside surface. The mobile phase was 0.1 mol/L of citrate buffer adjusted to pH 3.1 with sodium hydroxide. The mobile phase further contained 0.21 mol/L of dimethyloctylamine and 0.60–0.95 mol/L of sodium octyl sulfate. Because of column aging, the concentration of sodium sulfate in the mobile phase had to be increased to maintain the retention of the analytes.

With this detection technique three methods of quantification can be applied. The quantity of the compounds can be measured by determination of (1) the area or the height of chromatographic peaks at a given voltage, (2) the area or the height of voltammetric peaks at the chromatographic maximum, or (3) the area of the chromavoltammetric peaks. By measurement of the voltammetric wave in the chromatographic peak maximum the following detection limits were determined: DOPA < 1.5 fmol, DOPAC < 0.54 fmol, 5-HTP < 0.11 fmol, and HIAA < 0.36 fmol. These values were also the upper limits of these substances in the cells. The investigation concentrated on three giant neurons (cell volume ≃ 1.2 nl) from the land snail (*Helix as-*

persa). The neurons were those labeled F1 (right parietal ganglion), E4 (visceral ganglion), and D2 (left parietal ganglion) according to the map in Reference 6.

The content of Tyr, DA, Trp, and 5-HT in individual cells was qualitatively determined. They ranged from the detection limit to less than 10 and hundreds of fmol in a cell. Individual cells were also compared on a qualitative basis. After determination of the components a reproducible number was found in the chromatogram. The cells had similar amounts of amino acids, but they exhibited differences in both their quantitative and qualitative composition. For example, in one type of cell (D2) dopamine and serotonin were not detectable but another type of cell (E4) contained these components on the fmol level. Tyrosine and tryptophan was present in similar concentration levels in all the three types of cells. In the chromatograms a number of reproducible, unidentified peaks were found. Relative magnitude of the peaks remained constant, and they seem to indicate the momentary state of the neurons at the time of analysis.

In this example capillary liquid chromatography has been shown to enable a study that is not possible by analysis of individual organs as a whole. In addition to the study of a single cell, microcolumn liquid chromatography also enables us to determine the composition of extracellular fluid [7]. For this purpose a dialysis cannula [8], which can be implanted directly in the organism, is used. Sampling by means of the dialysis unit enables us to study changes in analyte composition over time, sometimes also under selected conditions. It is possible to study both substances originating in the organism and those carried to the organism from the outside, such as drugs [9]. The analyte is separated from proteins through a membrane with a molecular mass exclusion of around 10,000. Although this is a diffusion process, passage of the analyte through the membrane with a molecular mass exclusion of around 10,000. Although this is a diffusion process, passage of the analyte through the membrane is slow and in systems with conventional columns the recovery is only 50%. In miniaturized systems, however, the recovery sometimes reaches up to 90% at a flow rate 0.1 μl/min and 37°C [8]. Here also the detection limits for neurotransmitters mea-

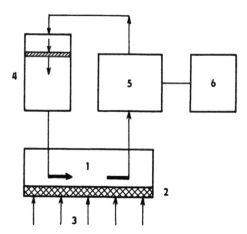

Figure 6.1 Dialysis device: (1) dialysis block; (2) membrane; (3) space for analyte withdrawal; (4) push-pull syringe device; (5) sampling device; (6) microcolumn liquid chromatograph (LC).

sured with an electrochemical detector were very low. For example, the detection limit for dopamine was 300 fg for the ratio S/N = 2.

The block diagram for on-line dialysis and microcolumn liquid chromatography is shown in Figure 6.1. Dialysis block 1 is provided with a suitable membrane 2 through which the dialysate diffuses from analyzed space 3. The dialysis device is washed with a liquid from push-pull syringe device 4. The sample is injected from injection device 5 to microcolumn liquid chromatograph 6 in intervals proportional to the flow rate of the liquid enriched by the analyte.

Probably the most frequently determined natural substances present in the organisms tested were catecholamines. Using direct sampling beyond the dialysis unit or classic sampling and treatment of a part of the organism, the changes in the content of catecholamines were determined for the tested animals (rat and cod) [10] under different conditions such as at rest, at work or under stress. In all these investigations electrochemical detection was applied to advantage. The sen-

sitivity of the analysis further increases with a reduction in column
diameter below 1 mm [11,12].

Attention has been paid to the determination of traces of amino
acids. Here also the two problems already described in detail occur:
the maximum sample volume that can be injected into the column
without loss of column efficiency and the sufficiently sensitive detection of substances that in general do not absorb light in the ultraviolet
(UV) region and do not exhibit electrochemical activity with current
electrodes.

The problem of increasing the detection sensitivity of amino acids
is often solved by precolumn or postcolumn derivatization (labeling).
For example, 5-dimethylaminonaphthalene-1-sulfonyl (DNS) amino
acids have a number of advantages in fluorimetric detection. With a
laser fluorimetric detector [13], a detection limit of 5×10^{-17} mol was
obtained. The column was 30 cm long, ID 0.5 mm, packed with
Silalsorb SPH-C18 ($d_P = 7$ μm). A sample chromatogram is given
in Figure 6.2. Another advantage of DNS derivatization is that the
absorption and fluorescence spectra are independent of the nature of
the amino acids. However, the spectra are partly dependent on the
mobile-phase composition. DNS amino acids can also be detected with
sufficient sensitivity amperometrically using a Pt electrode, as shown
in Figure 3.15. Considerable attention has been paid to derivatization
of amino acids [14]. Naphthalene-2,3-dicarboxaldehyde–labeled amino
acids [15] are electrochemically detected [2] in capillary liquid chromatography. A total of 18 amino acids can be detected and quantified
with detection limits in the atommolar range. The analysis technique
corresponds to present trends in the analysis of proteins. It enables us
to determine the amino acid composition of ever smaller quantities of
protein.

Efficient separation of a greater number of amino acids is always
connected with the gradient technique. The pH gradient, frequently
used in the past, is at present often substituted by the hydrophobic
component gradient (e.g., acetonitrile) in the buffer. The gradient
technique in trace analysis also enables injection of larger sample
volumes into the column without loss of column separation efficiency.

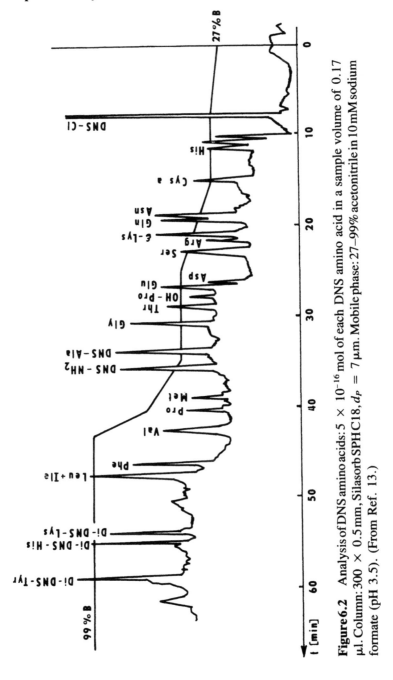

Figure 6.2 Analysis of DNS amino acids: 5×10^{-16} mol of each DNS amino acid in a sample volume of 0.17 μl. Column: 300×0.5 mm, Silasorb SPH C18, $d_P = 7$ μm. Mobile phase: 27–99% acetonitrile in 10 mM sodium formate (pH 3.5). (From Ref. 13.)

This again positions the high mass sensitivity of the analysis side by side with its high concentration sensitivity.

In many cases injection of the pH-generated gradient [16] may prove successful. This technique enables separation and focus of peaks of ionizable substances: anions, cations, and ampholytes. The method consists of establishing the phase equilibrium of several (three to six) salts buffering in the required pH range and having suitable pK_a values over the sorbent, whose sorption properties are controlled by dynamic modification. The equilibrium established in the column is then destroyed by injection of the sample dissolved in a solvent of suitable pH. The sample pulse causes not only a change in the equilibrium pH but simultaneously also a change in the charge of the ionizable groups of the stationary phase. The ion analytes are then retained and focused. Their elution is controlled by migration of the pulse of the change in phase equilibrium in the buffering salts present in the mobile phase.

An example of ampholytic solute elution enables us to explain the mechanism of the sample-induced internal pH gradient. If ampholytes occur in an acid medium, that is, if they are dissolved in a solvent of sufficiently low pH, they have the form of cations. If the gradient changes from low values of pH to higher values, the stationary-phase charge changes from negative to positive depending on the character of active groups on the surface. For example, sulfonic groups on the surface cannot act as a buffer. Nevertheless, the residue silanol groups on the surface carrying the reversed phase can do so very well. An ampholyte sample in acid medium is injected into the top of the column. The cationic form of an ampholyte is strongly sorbed onto the surface and is changed to the anionic form as a result of sample injection. During elution of the injected sample pulse the pH increases, the ampholytes are changed to their anionic forms, and they are excluded from the surface of the sorbent containing sorbed sulfonic groups. An example of ampholyte separation [16] is given in Figure 6.3. Detection of the analytes was performed by a spectrophotometric detector. For detection of the pH course and conductivity, a bifunctional detector [17] connected beyond the spectrophotometric detector was used. This method can be used both for the analysis of cationic substances and for the analysis of anions. Modification of the method consists of

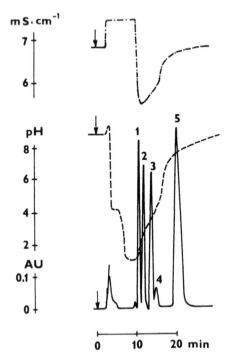

Figure 6.3 Separation of ampholytes by microcolumn LC with a sample-induced internal pH gradient. Microcolumn: 150 × 1 mm, SGX C18 d_P 5 μm sorbent (Tessek Progue, Czechoslovakia). Sample: mixture of ampholytes in 60 μl phosphoric acid, 0.5 mol/L. Solutes: (1) nicotinic acid; (2) 4-aminosalicylic acid; (3) 4-aminobenzoic acid; (4) 3-aminobenzoic acid; (5) 4-nitro-2-aminophenol. Detection: UV spectrophotometer with 0.5 μl flow cell, λ = 260 nm, conductivity and pH-simultaneous conductivity and pH detector with 0.15 μl flow cell. Mobile phase: aqueous solution of sodium salts (concentrations in mmol/L: sulfate (20), chloroacetate (20), formate (10), acetate (2), MES (1), phosphate (1), and dodecylsulfate (0.4) adjusted to pH 8.5 by NaOH. (From Ref. 16.)

selection of a suitable stationary phase or its modifier, selection of a system of buffer salts in the mobile phase, and selection of the solvent pH in which the sample is dissolved.

Another topical analytic problem is separation, identification, and determination of biomacromolecules. The requirements for determining the components of complex mixtures in concentrations of around 10^{-9} g/g. Attention is given mainly to active proteins and peptides. The most frequently used techniques are liquid chromatography and electrophoresis. In both cases the miniaturized version of the techniques is used to advantage [14].

Size-exclusion chromatography provides theoretically predictable conditions for the separation of macromolecules. The highest selectivity is obtained by affinity chromatography. Ion-exchange chromatography exhibits the highest sorption capacity together with the possibility of refining biologic materials. The disadvantage of these systems is the relatively low efficiency of the columns. In contrast, hydrophobic interactions in the reversed phase also exhibit the highest efficiencies for biologic macromolecules. A disadvantage—the loss of the biologic activity of these materials—can be overcome by using suitably selected buffers.

6.2 PHARMACEUTICAL SAMPLES

Liquid chromatography has become an integral part of the pharmaceutical industry. It is used not only for checking the quality of the manufactured pharmaceuticals but also for the study of pharmacokinetics, the time stability of pharmaceuticals, and the like. Attention is paid to racemates [18] whose separation has been developed on the basis of the high selectivity of suitable stationary phases. An example using β-cyclodextrin as the stationary phase [19,20] is the use of microcolumns for separation of the optical isomers of amino acids (Fig. 6.4).

Highly selective phases, such as β-cyclodextrin, can also be applied for separation of pharmaceuticals used for their activity in suppressing tumors. The *cis*-diaminodichloroplatinum(II) (CDDP) com-

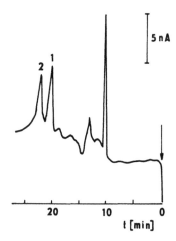

Figure 6.4 DNS of derivatives of optical isomers of amino acids. Column: 0.2 × 200 mm, β-cyclodextrin bound to silica gel carrier, d_P = 5 μm. Mobile phase: acetonitrile-water 15:85 (vol/vol) + 3 × 10^{-3} mol/L of $NaNO_3$. Detector: amperometric, polarization voltage 1.5 V, Pt electrode. Sample: (1) dansyl-l-phenylalanine; (2) dansyl-d-phenylalanine.

plex is widely used for treatment mainly of solid tumors. With the increasing content of acetonitrile in the mobile phase the retention of CDDP in the column with a β-cyclodextrin stationary phase increases [2]. An example of this analysis making use of an electrochemical detector is given in Figure 6.5. With the electrochemical detector in the oxidation mode the minimum detectable amount of CDDP, 1.5 × 10^{-11} mol, was determined and the minimum detectable concentration in the sample was 5 ppm (5 μg/ml).

Microcolumns coupled with amperometric detection can be used for a number of pharmaceutical substances [22]. A mixture of sulfanilic acid, sulfanilamide, sulfacetamide, sulfathiazole, and sulfadimidine can be separated on a 15 cm long column with a reversed phase. An example is given in Figure 6.6. The chromatogram shows not only good and quick separation of the components but also good sensitivity of the analysis. With the amperometric detector minimum detectable concentrations can be obtained in units of μg/L. Electrochemical de-

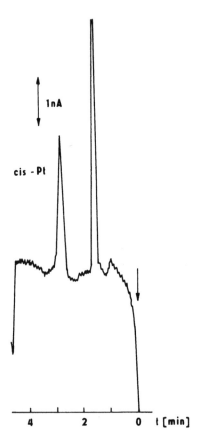

Figure 6.5 Determination of cytostatic platinum in body fluid. Column: 0.5 × 60 mm, β-cyclodextrin bound to silica gel carrier, $d_P = 5$ μm. Mobile phase: 95% acetonitrile, 0.5 mol/L of $NaNO_3$, flow rate 8.5 μm/min. Detector: see Figure 6.4. Sample: 0.1 μl CDDP in body liquid (61 ng/ml).

tection can also be used for tetracyclines. Their detectability is conditional on the presence of the phenolic group in their structure. An example applying the reversed phase in the column is given in Figure 6.7. Chromatography with direct phases can be used for separation and determination of structurally very close derivatives of phenothiazine and thioxanthene. The organic mobile phase, acetonitrile, is in this case modified by a concentrated aqueous solution of ammonia.

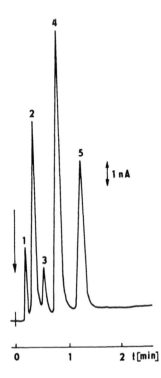

Figure 6.6 Separation of sulfonamides. Column: 0.7 × 150 mm, Lichrosorb RP 18, d_P = 7 μm. Mobile phase: acetonitrile-water 3:97 (vol/vol) + 0.1 mol/L NaC10$_4$, velocity 10.7 mm/s. Detector: amperometric; see Figure 6.4. Sample (0.2 μl): (1) sulfanilic acid (16 ng); (2) sulfanilamide (11 ng); (3) sulfacetamide (8.8 ng); (4) sulfathiazole (24 ng); (5) sulfadimidine (78 ng).

The electrochemical detection is based on oxidation of delocalized π electrons. The detection is easy and sensitive. An example is given in Figure 6.8. Analyses that are frequently used for checking food products are separations of the derivatives of 4-hydroxybenzoic acid. This separation can be carried out on reversed phase, and a sample chromatogram is shown in Figure 6.9. The sensitivity of amperometric detection can be compared with that of spectrophotometric detection in the UV region. Vitamins are frequently analyzed in biochemical practice. An example is given in Figure 6.10.

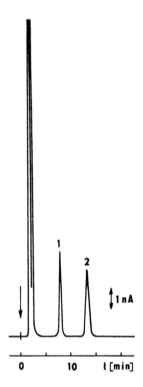

Figure 6.7 Separation of tetracyclines. Column: see Figure 6.6 Mobile phase: acetonitrile-water 25:75 + 10^{-3} mol/L of $NaClO_4$. Sample (0.2 μl): (1) rolitetracycline (21.4 ng); (2) tetracycline (30.6 ng).

6.3 INORGANIC ANALYSIS

Microcolumns are convenient for separation and determination of trace concentrations of inorganic ions. Three basic chromatographic systems are used for separation of metal cations: (1) separation of cations on a solid cation resin, (2) separation of complexes of the corresponding cations on reversed-phase columns, (3) separation on solid sorbents prepared by chemical bonding of a complexing agent.

   If the cation resin is used for separation of cations by an ion-exchanging mechanism, the mobile phase contains a mineral acid and

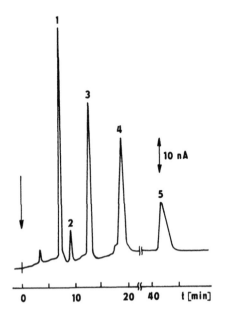

Figure 6.8 Chromatogram of phenothiazine mixture. Column: 0.7 × 150 mm, Lichrosorb SI 100, d_P = 5 μm. Mobile phase: acetonitrile + 10^{-3} mol/L of NaClO$_4$ + 10^{-2} mol/L of NH$_3$, velocity 0.75 mm/s. Sample (0.2 μl): (1) levopromazine (22.8 ng); (2) chlorportixen (41.6 ng); (3) chlorpromazine (18.4 ng); (4) thioridazine (20.4 ng); (5) prochloroperazine (19.2 ng).

a competitive cation. It is also possible to add the complexing agent, most often hydroxycarboxylic acid, to the mobile phase [23]. Complexes (negative or neutral) in equilibrium with cations occur in the mobile phase. Retention of the cations is determined by the stability of the complexes. The mobile phase often also contains the competitive cation.

Using conductometric detection, minimum detectable concentrations of 50 ppb Mg^{2+} and 80 ppb Ca^{2+} [23], for example, were obtained. The minimum detectable concentrations of dithiocarbamate complexes of transition metals measured on the amperometric detector anode [24] were in the range from 0.01 to 1 ppm. Atomic absorption

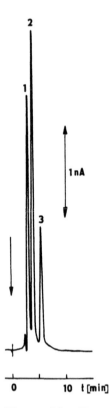

Figure 6.9 Chromatogram of derivatives of *p*-hydroxybenzoic acid. Column: see Figure 6.3. Mobile phase: acetonitrile-water 60:40 (vol/vol) + 0.1 mol/L of $NaClO_4$ + 10^{-3} mol/L of $HClO_4$. Sample (0.2 µl): (1) 4-hydroxybenzioc acid (9.3 ng); (2) methyl ester of 4-hydroxybenzioc acid (11.2 ng); (3) propyl ester of 4-hydroxybenzoic acid.

detection [25] reached a minimum detectable concentration from 0.1 to 10 ppm. When detecting metal complexes by spectrometric detection in the UV or visible regions of the spectrum, the minimum detectable concentrations [26] were in the range from 0.01 to 1 ppm. Minimum detectable concentrations in the low range of ppb were obtained with a reaction detector.

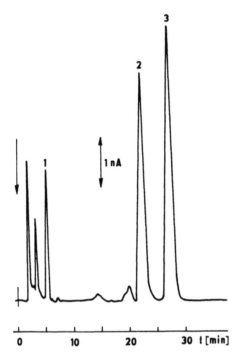

Figure 6.10 Chromatogram of some vitamins. Column: see Figure 6.3. Mobile phase: acetonitrile + 0.1 mol/L of NaClO$_4$. Velocity 1.5 mm/s. Sample (0.2 µl): (1) A-acetate (5.6 ng); (2) D$_2$ (36.4 ng); (3) E (710 ng).

The combination of catex as a strong stationary phase with a complexing agent to regulate the strength of the mobile phase and a competing cation added to the mobile phase enables efficient enrichment of the analyzed cations on the microcolumn [27]. The trace analysis of some bivalent cations was performed on a column of inside diameter 0.5 mm, 150 mm long, packed with the cation resin Silasorb S, particle diameter 7.5 µm (Lachema, Brno, Czechoslovakia). The mobile phase was treated by dissolving tartaric acid and ethylenediamine in redistilled water. The sample was injected by way of a four-port injection valve with a 0.2 µl internal loop (Valco Instruments Houston, TX) and by a six-port injection valve with an external loop of 100 µl and

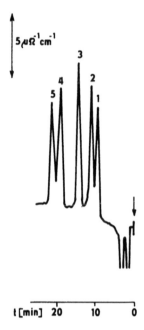

Figure 6.11 Separation of cations of transition metals. Column: fused silica 0.25 × 150 mm, Silasorb S, $d_P = 7.5$ μm. Mobile phase: 4.5×10^{-3} M ethylenediamine; and 3×10^{-3} M tartaric acid, pH 4. Injection: 0.5 μl. Detector: conductivity detector. Sample (1) Zn^{2+} (81.7 ng); (2) Ni^{2+} (146.7 ng); (3) Co^{2+} (147.5 ng); (4) Cd^{2+} (352.8 ng); (5) Mn^{2+} (171.8 ng).

1 ml volume. The chromatograph was equipped with a conductometric detector (Laboratory Instruments, Prague, Czechoslovakia). The retention of cations depends on the concentration of tartaric acid and that of ethylenediamine. The mobile-phase pH also plays an important role. It is recommended to work with a pH in the range from 3 to 5. At pH values < 3 tartaric acid loses its capacity to form complexes; at pH > 5 protonation of ethylenediamine is incomplete.

Examples of the analysis of cations are given in Figure 6.11. In the separation of transition metal cations the injections were 0.2 μl. Conductometrically detectable quanta were of the order of magnitude of tens of ng. In the examples in Figure 6.12, 1 ml sample was injected

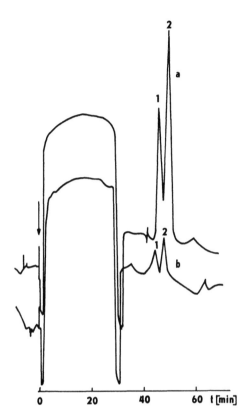

Figure 6.12 Determination of Ca and Mg in distilled (a) and redistilled (b) water. Column: 0.7 × 150 mm, Silasorb S. Mobile phase: tartaric acid 3.0 × 10^{-3} mol/L and ethylenediamine 4.5 × 10^{-3} mol/L, pH 4: (1) Mg; (2) Ca.

and the minimum detectable concentrations ranged from 10^{-8} to 10^{-9} mol/L.

The reversed phases are also applicable to the analysis of ions. An example of the analysis of anions is given in Figure 6.13. The detection limit from one-tenth to 10 ng can be obtained with a column of ID 0.7, 150 mm long, with a conductometric detector. For separation of anions a dynamically prepared ion exchanger was applied. Tetrabu-

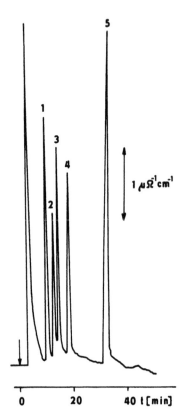

Figure 6.13 Chromatogram of inorganic anions. Column: 0.7×150 mm, Silasorb SPH C18, $d_P = 7.5$ μm. Mobile phase: 0.75 mmol/L of tetrabutylammonium phthalate and 0.25 mmol/L of potassium phthalate, pH 6.3. Detector: conductivity detector. Solutes: (1) Cl^- (35 ng); (2) NO_2^- (46 ng); (3) Br^- (80 ng); (4) NO_3^- (62 ng); (5) SO_4^{2-} (96 ng).

tylammonium phthalate with potassium phthalate was used in the mobile phase. The selectivity and sorption capacity of the column can be regulated by ionogenic component concentrations of the order of units to tenths of units of mmol/L.

REFERENCES

1. Kenedy R. T., Jargenson J. W.: Anal. Chem. 61, 436 (1989).
2. Knecht L. A., Guthrie E. J., Jorgenson J. W.: Anal Chem. 56, 479 (1984).
3. St. Claire R. L., Jorgenson J. W.: J. Chromatogr. Sci. 23, 186 (1985).
4. White J. G., St. Claire R. L., Jorgenson J. W.: Anal. Chem. 58, 293 (1986).
5. White J. G., Jorgenson J. W., Anal. Chem. 58, 2991 (1986).
6. Kerkut G. A., Lambert J. D. C., Gayton D., Walker R.: J. Comp. Biochem. Physiol. [A] 50A, 1–25 (1975).
7. Church W. H., Justice J. B. Jr., in: Advances in Chromatography, Vol. 28 (eds.), J. C. Giddings, E. Grushka, P. R. Brown, Marcel Dekker, New York, 1989, p. 165.
8. Wages S. A., Church W. H., Justice J. B. Jr.: Anal. Chem. 58, 1649 (1986).
9. Tjaden V. R., De Bruijn E. A., Van Der Greet J., Lingeman H.: LC-GC 6, 600 (1900).
10. Ehrenström F.: Life Sci. 43, 615 (1988).
11. Šlais K., Krejčí M., Chmelíková. J., Kouřilová D.: J. Chromatogr. 388, 179 (1987).
12. Krejčí M., Šlais K., Kunath A.: Chromatographia 22, 311 (1986).
13. Lobasov A. Ph., Mostovnikov V. A., Nechaev S. V., Belenskij B. G., Kever J. S., Korolyova E. M., Maltsev V. G.: J. Chromatogr. 365, 321 (1986).
14. Novotny M.: J. Microcolumn Separation 2, 7 (1990).
15. Oates M. D., Jorgenson J. W.: Anal. Chem. 61, 432 (1989).
16. Šlais K.: J. Microcol. Sep. 3, 191 (1991).
17. Šlais K.: J. Chromatogr. 540, 41 (1991).
18. Jira T., Vogt C., Beyrich T.: Pharmazie 43, 385 (1988).
19. Krýsl S., Smolková-Keulemansová E.: Chem. Listy 79, 919 (1985).
20. Vaisar T., Vaněk T., Smolková-Keulemansová E.: Czech Patent Appl. PV-7640-88.
21. Doležel P., Kouřilová D., Krejčí M., Smolková-Keulemansová E.: J. Microcolumn Sep. 2, 241 (1990).
22. Vespalcová M., Šlais K., Kouřilová D., Krejčí M.: Českoslov. Farm. 33, 287 (1984).
23. Sevenich G. J., Fritz J. S.: Anal. Chem. 55, 12 (1983).

24. A. M. Bond, Wallace G. G.: Anal. Chem. 56, 2085 (1984).
25. Maketon S., Otterson E. S., Tarter J. G.: J. Chromatogr. 368, 395 (1986).
26. Schwedt G.: Chromatographia 12, 613 (1979).
27. Kouřilová D., Nguyen Thi Phuong Thao, Krejčí M.: Int. J. Environ. Anal. Chem. 31, 183 (1987).

7
Combination of Microcolumn Liquid Chromatography with Spectral Identification Methods

Microcolumn liquid chromatography, with its extraordinary separation efficiency, is ever more frequently used for the separation of complex multicomponent mixtures. Mobile phases and sometimes also the sample matrix intensively influence retention, which, together with the relatively low reproducibility of sorption properties of the sorbent or the stationary phase fixed on the capillary column wall, causes relatively low reproducibility of retention data of the complex mixture components. Also, in many cases standards of individual components are not available. The amount of analyte injected into microcolumns and capillary columns eliminates the preparation and independent treatment of the analytes. All these circumstances as well as a number of other problems resulting from the requirements of trace analysis or analysis of totally unknown samples led to attempts to increase the identification possibilities of chromatographic analysis.

The most efficient combinations for liquid chromatography (LC) proved to be those with mass spectrometry (MS) and with spectrometry in the ultraviolet and visible regions of the spectrum (UV/VIS) and in the infrared region of the spectrum (IR). This last combination is used at present exclusively in connection with Fourier transformation of spectra (FTIR). Most attention in microcolumn chromatography is paid to the combination LC-MS, for several reasons. The spectra are highly conclusive, and the detection technique allows are to work with relatively small time constants, which prevents deterioration of separation in the mass spectrometer. The mass spectra can be scanned with sufficient speed, which means that during one peak several hundred spectra can be scanned. The speed of spectra scanning also enables one to distinguish overlapped peaks in the chromatogram.

 ⁎ The definitive spectrometric methods FTIR and UV/VIS are in a rapidly moving stage of apparatus development and application. The combination of microcolumn chromatography with spectral methods has a number of additional advantages. It eliminates damaging of the sample between individual analytic steps, and the trace analysis contamination of the sample by the laboratory environment or personnel is prevented. Simultaneously it decreases danger when working with toxic substances. Mass spectrometry further allows quantitative measurement with application of isotopically traced substances.

7.1 COMBINATION WITH MASS SPECTROMETRY

In the late 1950s, when the first combination of a gas chromatograph and a mass spectrometer [1] appeared, the barrier between the atmospheric pressure at the column outlet and vacuum in the active parts of the mass spectrometer was overcome. The combination of liquid chromatography with mass spectrometry brought other serious problems connected with the properties of the liquid phase flowing out of the column [2–6]. If the usual relationship for calculation of pressure p is used,

$$p = \frac{F_m \, \rho \, V_{\text{mol}} \, p_0}{MF_P} \tag{7.1}$$

where F_m is the flow rate of the mobile phase entering the spectrometer (ml/s), ρ is the mobile-phase density (g/ml), V_{mol} is the gas molar volume (22.4 L/mol), p_0 is the atmospheric pressure (Pa), M is the molecular mass of the effluent flowing out of the column (g/mol), and F_P is the exhausted gas flow rate (L/s).

Assuming that the mobile phase is methanol ($M = 32$ g/mol and $\rho = 0.8$ g/cm^{-3}), the pumps of the mass spectrometer pump off $F_P = 150$ L/s, and all the effluent from the column is transferred to the spectrometer, we find that at the usual flow rate in liquid chromatographs with conventional columns, $F_m = 1$ ml/min and the pressure $p = 6.25$ Pa; at the usual flow rate for microcolumns and $F_m = 10$ or 100 μl/min, $p = 6.25 \times 10^{-2}$ Pa or 6.25×10^{-1} Pa. Assuming that the highest permissible pressure in the spectrometer should be about 10^{-1} Pa, we can see that for effluent injection to the column flow rates of 10 μl/min and lower are suitable. These values are suitable for work with packed fused silica microcolumns ($d_c = 0.1$–0.3 mm) or with capillary columns ($d_c < 50$ μm). This and the fact that the usual analyte injections ($\simeq 1$ μg) into the column agree well with the detection possibilities of the mass spectrometer demonstrate the prospects for further development and application of the combination of microcolumn chromatographs with mass spectrometers.

Mass spectrometry is based on the generation of ions from analyte molecules and their subsequent separation. For generation of ions one of four methods is most frequently used at present: (1) electron ionization (EI), (2) chemical ionization (CI), (3) thermospray ionization [7] (TS), and (4) ionization by fast atom bombardment [8] (FAB) or secondary ion mass spectrometry (SIMS).

Electron ionization mass spectrometry (EIMS) provides a rich spectrum of molecular fragments. The protonized fragment, that is, the pseudomolecule ion $(M + H)^+$, is usually weak; sometimes it is not present in the spectrum at all. In EIMS the sample is introduced in the gas phase to the ionization source, where it is bombarded by electrons. Chemical ionization is a gentler method that provides mostly the protonized fragment corresponding to the analyte molecular mass, $(M + H)^+$, that is, the pseudomolecular fragment. In CIMS the analyte molecules in the ionization source of the mass spectrometer are surrounded with an excess of reaction gas (in the source there is usually a relatively high pressure, such as 1.33×10^2 Pa), such as methane or ammonia. This gas is ionized by the same technique as in EI. The ionized molecules react among themselves and simultaneously transfer protons to the analyte molecules, causing generation of the fragment $(M + H)$.

Ionization by thermospray or electrospray is suitable for thermally unstable substances. This is an even more careful method than CI. The analyte is ionized as a result of the charge present on the spray droplets. The effluent must contain ionizable and volatile substances, such as ammonia, formic acid, trifluoroacetic acid, or ammonia acetate damping solution. The ammonia and acetate ions are irregularly distributed in droplets, and consequently the droplets are charged. This method of ionization leads to the origin of mostly pseudomolecular fragments $(M + H)^+$ and $(M + NH_4)^+$.

FABMS provides both pseudomolecular and ionic fragments. It is used for polar and thermally labile analytes. In contrast to the SIMS method, in which the solid sample is placed in the ionization source and bombarded with Cs^+ ions, for example, in FAB the sample is mixed with a viscous liquid (glycerol) and the mixture is bombarded with fast atoms, for example xenon.

The important part of the LC-MS combination is the connection (interface) of the liquid chromatograph with the mass spectrometer. At present two categories of such devices exist based on mechanical interface and spraying.

The mechanical LC-MS interface makes use of moving conveyers, most frequently polyimide tapes. These carry the effluent from the column to the chamber in which the mobile phase is evaporated; they also carry the analyte to the ionization source, where part of it is ionized and the rest is carried to the purifying and drying chamber. Bushings that enable the conveyer to pass through the ionization source must be technically perfect. They are usually equipped with double or triple seals to ensure low pressure in the ionization source of the mass spectrometer. Even though this is the oldest technique of transport of the analyte to the ionization source [9], derived from the transport detection with a ionization sensor [10], at present it seems to be more reliable in combination with conventional columns than with micro-columns. Analyte transport techniques are further developed. For off-line sample transport to the SIMS system electrospray application was proposed. This relatively simple technical solution [11] consists of applying voltage at the microcolumn outlet. The microcolumn ends in a capillary about 100 μm in diameter to which a voltage from 0 to 2 kV is applied. A rotating disk with a suitable foil is placed against the capillary. The disk is supplied with voltage from −3 to −6 kV. The nonvolatile solutes leaving the chromatographic microcolumn form a chromatogram on the foil. The foil is placed in the mass spectrometer, and from different parts of the foil mass spectra of the corresponding analytes are recorded.

The advantage of the application of microcolumns in combination with mass spectrometer become evident especially when the effluent directly enters the ion source. The effluent is not simply evaporated at the column outlet. Evaporation causes cooling of the column outlet. In most cases the mobile phase is evaporated in the column, where the effluent nonvolatile components are accumulated, which very often leads to an increase in the column hydraulic resistance, sometimes even to choking. Therefore, the effluent from the column is transported to the ion source of the mass spectrometer in the form of an aerosol. The chromatographic column must be separated from the evacuated spaces of the mass spectrometer, most often with fritted glass or a capillary. These elements create conditions for spray generation and also for the generation of droplets of suitable dimensions. The droplet diameter usually corresponds to double the diameter of the jet or capillary through which the liquid flows out. The droplets are formed at a certain minimum velocity of the liquid streaming through the jet. This minimum velocity is a function of the surface tension, the density of the flowing liquid, and the jet diameter. The diameter of the capillaries is usually expressed as less than 10 μm. The fritted glass must have similar diameters. Its thickness ranges around tens of μm.

The sprays used for the transport of the effluent to the mass spectrometer either merely transport the liquid to the spectrometer [12,13] or transport is accompanied by effluent ionization [14]. The systems enabling both ionization and desolvation of the analyte are considered particularly advantageous. In the process of spraying the analyte is enriched and the mass spectrum is simplified because of the absence of mobile-phase ions. The mobile phase can provide quite a complex spectrum, which in chemical ionization can be more intensive than the analyte spectrum. For example [15], methanol provides 24 molecular fragments. The most important fragment is $(M + H)^+$, but a further 9 fragments provide a signal higher than 25% of the basic fragment. These signals affect the analysis, and they can be removed from the spectrum during computer processing. Such data processing, however, always leads to an increase in the minimum detectable analyte concentration. Therefore, it is of greater advantage to remove the solvent from the analyte or at least increase the analyte concentration in the solvent.

The atomized particles, droplets, arising from liquid spraying are charged. The charge surface density and the mechanism of its origin are different, and they depend on the method of spraying. For example, the origin of charged droplets is totally different for thermospray and electrospray. In the former most of the liquid in the heated capillary changes into gas, the flow velocity increases, and the droplets leave the capillary at supersonic speed. Statistical fluctuations in the number of anions and cations lead to the formation of a positive or negative charge on each droplet. The charged flow of particles enters the first vacuum stage of the mass spectrometer. Similarly to thermospray, spray based on effluent atomization by gas (MAGIC) also provides charged particles, droplets, whose charge is influenced by the potentials of the electrode placed close to the spraying jet. In both cases the charge is a result of droplet formation. At the electrospray [14] the droplets are formed because of charging of the liquid leaving the column. The liquid is conducted to the spraying space by a capillary whose end is provided with a voltage of several kV with respect to the spraying chamber walls. The charge is transferred to the liquid. This leads to coulombic repulsive powers that gradually overcome the surface tension, and the liquid is dispersed in fine charged droplets. Mobile-phase evaporation leads to an increase in the charge surface density and a decrease in the radius of curvature of the droplets. Consequently, the electrical field intensity increases up to the value at which the ions originate. Therefore, electrospray can be used directly as the ion source. As such the technique serves well for processing of liquid flow rates of the order of μl/min. This LC-MS combination is suitable for both microcolumn and capillary liquid chromatography.

In applying electrospray as the ionization source, beyond the column outlet a chamber with a gas bath can be placed where the solvent is evaporated. Beyond the chamber a system of ion optics is placed by which the ion stream is focused to the mass spectrometer. Under such circumstances the solute is enriched and in the mass spectrometer the mobile-phase ions are considerably suppressed.

In some cases electrospray is used only for atomization of the liquid, and the total effluent from the chromatographic column is conducted to the ionization source of the mass spectrometer [16]. A schematic representation of such a device is provided in Figure 7.1A; the important element of the electrospraying device is the liquid outlet. The ratio of the outside and inside jet diameters is critical. It should be as close to 1 as possible, which means that the thickness of the capillary wall should be minimal. Examples of the technical solutions to this problem are given in Figure 7.1B. An electrode is placed in the immediate proximity of the jet. With a metal jet the atomizing voltage can be conducted directly to the jet.

Even if the basic demand on the LC-MS combination is the increase in identification capacity of the chromatographic analysis, the mass spectrometer can also be used as a universal detector. In most cases the mass spectrometer does not cause peak spreading in microcolumn chromatography—sometimes not even in capillary liquid chromatography. Nevertheless, the most important function of the combination is reliable identification of the analyzed substances. Figure 7.2 compares the mass spectrum obtained in a chromatographic experiment with a packed microcolumn of inside diameter 0.2 mm and a spectrum from the spectrum library [13]. The agreement of both spectra is very good. Miniaturization of the chromatograph even when combining liquid chromatography and mass spectrometry increases the range of application of these techniques.

7.2   COMBINATION WITH INFRARED SPECTROMETRY

The high identification capacity of infrared spectroscopy inspired the effort to scan the spectra of individual chromatographically separated

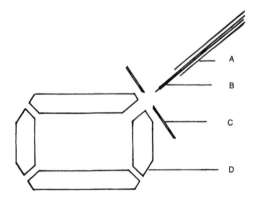

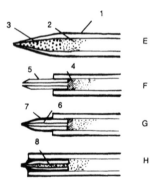

Figure 7.1 Electrospray as the inlet to the ionization source of a mass spectrometer according to Reference 16. (A) Earthed inlet capillary; (B) fused silica microcolumn; (C) electrode, 2–3 kV; (D) ionization source; (E–H) types of endings of fused silica columns: (1) fused silica microcolumn; (2) chromatographic packing; (3) coarse-grained sorbent (d_P = 40 μm); (4) sintered glass; (5) fused silica capillary with inside diameter 50 μm; (6) fused silica jet drawn to a diameter of 50 μm; (7) heat-resistant epoxy resin; (8) sorbent (d_P = 10 μm)

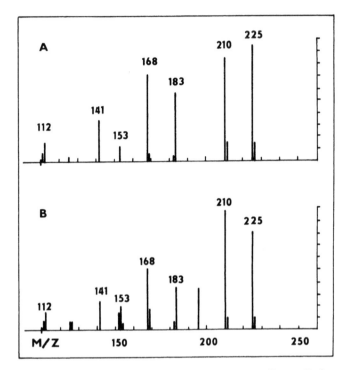

Figure 7.2   Comparison of mass spectra according to Reference 13. (A) LC-MS combination; (B) NBS library. Sample was Prometon [2-methoxy-4,6-bis(isopropylamino)-1,3,5-triazine].

analytes. The high scanning velocity of modern infrared spectroscopy with Fourier transformation (FTIR) created good conditions for this effort. First, the connection with gas chromatography [17] was made. The LC-FTIR connection gave rise to a number of complications as a result of properties of the mobile phases since most of them absorb in the infrared region of the spectrum. Therefore, it is necessary to look for "windows" in the spectrum that can be used for analysis. Detectors working on a single wavelength in the infrared region are less often applied in liquid chromatography. For the micro-LC-FTIR combination two basic methods are used. The first makes use of the possibility of subtracting the mobile-phase spectrum from the final spectrum during data processing. This step is enabled by computer evaluation, but it is connected, as in LC-MS, with the loss of method sensitivity. The other method consists of removing the mobile phase before scanning the spectrum, which leads to an increase in both the method sensitivity and the reproducibility of the scanned spectra.

In LC-FTIR two systems are used:

1. Flow rate cells allow one to scan spectra during the chromatographic process. Regions of the infrared spectrum must be selected in which mobile-phase absorption is sufficiently small [18–20].
2. Transport systems enable one to remove the mobile phase before spectrum scanning or at least to substitute the mobile phase with a liquid suitable for analyte spectrum scanning [21–23].

The first method is the simplest for infrared spectrometry. It enables one to make use of the spectrometer as a real chromatographic detector. Its application, however, is limited exclusively to those regions of the spectrum in which the mobile phase does not absorb, which leads to the loss of information, or it is necessary to design the cell with a sufficiently short optical path that the mobile-phase spectrum can be suppressed. The results are then connected with a loss of considerable sensitivity. The last technique can be used, for example, with a cell [24] in which the effluent from the column flows along the reflection wall. With respect to the very short optical path, the spectra of the solutes can also be scanned using highly absorbing aqueous mobile phases. However, this also causes the sensitivity to decrease intensively.

With conveyers diffusive reflection spectroscopy and transmission spectroscopy are used. For diffusive reflection FTIR a conveyer [22,25] consisting of a series of containers (vessels containing powdered KBr) was designed. The defined effluent volume flows out of the column into individual vessels. After evaporation of the solvent the spectra are scanned. This conveyer, however, does not allow a continual record of the chromatogram, and all the dangers that follow from the analysis of effluent discrete fractions are implied. Some components can be overlooked, and sometimes one container may contain more components because two peaks immediately follow one another in the chromatogram.

For transmission FTIR spectroscopy a conveyer [26] based on a rotating KBr crystal was designed. After mobile-phase evaporation the crystal is automatically shifted to the path of the passing beam. Thus the spectra can be scanned continuously and the spectrometer can be used as the chromatographic detector.

Both these conveyers proved successful only for microcolumns with direct phases. All the spectroscopic systems containing potassium bromide, in both powder and crystal line form, can work only with nonaqueous mobile phases, and therefore they are not suitable for work with reversed phases. In such a case water must be removed from the effluent before it enters the spectrometer. A device similar to a chain conveyer used for LC-MS or for the combination of liquid chromatography with a flame ionization detector [9,10] was applied. In fact, the mobile phase is removed and the solute is extracted by the liquid, which can be used in the IR spectrometer.

From the point of view of wider application, sensitivity, and good reproducibility of the spectra, a conveyer [23,27] that uses an aluminum reflection disk is of advantage. The effluent from the column sprayed with gas falls onto the disk. At the constant velocity of the disk rotation a spiral trace corresponding to a chromatogram remains on the disk reflection area after evaporation of the liquid. The disk is transferred to the reflection FTIR spectrometer, and the spectra are scanned at short time intervals. At the constant velocity of the disk rotation the chromatogram can be reliably reconstructed and a sufficient number of spectra can be scanned from each peak so that the information can be processed statistically. This conveyer works reliably with both direct and reversed phases.

Even though a number of authors have proposed the application of LC-FTIR with conventional columns, the practical combination in analytic laboratories seems to be mainly in connection with microcolumns. The increased concentration of the solute at the column outlet makes mobile-phase removal easier; in some cases it also reduces the negative influence of the mobile phase on the spectrometer response. Good reproducibility of the spectra, especially their correspondence with statistical and library spectra, make the micro-LC-FTIR combination one of the most powerful identification methods in the analytic chemistry of organic substances.

REFERENCES

1. Halmes F. C., Marrell F. A.: Appl. Spectrometry 11, 86 (1957).
2. Vouros P., Karger B. L.: New Methods Drug Res. 1, 45 (1985).
3. Niessen W. M. A.: Chromatographia 21, 277 (1986).
4. Niessen W. M. A.: Chromatographia 21, 342 (1986).
5. Henion J. D., in: Microcolumn HPLC (P. Kucera, ed.) J. Chromatogr. Library, Vol. 28, Elsevier, Amsterdam, 1984.
6. Stroh G. J., Rinehart K. L.: LC-GC 5, 562 (1987).
7. Blakley C. R., Carmody J. J., Vestal M. L.: Anal. Chem. 52, 1636 (1980).
8. Barber M., Bordoli R. S., Sedgwick R. D., Tyler A. N.: J. Chem. Soc. Chem. Commun. 325 (1981).
9. McFadden W. H., Schwartz H. L., Ewans S.: J. Chromatogr. 122, 389 (1976).
10. Scott R. P. W., Scott G. G., Munrone M., Hess J. Jr.: J. Chromatogr. 99, 395 (1974).
11. Beavis R. C., Bolbach G., Ens W., Main D. E., Schneler B., Standing K. G.: J. Chromatogr. 359, 489 (1986).
12. Baldwin M. A., McLafferty F. W.: Org. Mass Spectrometry 7, 111 (1973).
13. Tsuda T., Keller G., Ston H. J.: Anal. Chem. 57, 2280 (1985).
14. Whitehouse G. M., Dreyer R. N., Yamashita M., Fenn J. B.: Anal. Chem. 57, 675 (1985).
15. Voyksner R. D., Hoas J. R., Bursey M. M.: Anal. Chem. 54, 2465 (1982).
16. Alborn H., Stenhagen G.: J. Chromatogr. 394, 35 (1987).
17. Griffiths P. R., de Haseth J. A., Azzaraga L. V.: Anal. Chem. 55, 1361A (1983).
18. Johnoson C. C., Hellegh J. W., Taylor L. T.: Anal. Chem. 57, 610 (1985).
19. Brown R. S., Taylor L. T.: Anal. Chem. 55, 723 (1983).

20. Jinno K., Fujimoto C., Vematsu G.: Int. Lab. 3, 48 (1983).
21. Conray C. M., Griffiths P. R., Duff P. J., Azzaraga L. V.: Anal. Chem. 56, 2636 (1984).
22. Conray C. M., Griffiths P. R., Jinno K.: Anal. Chem. 57, 822 (1985).
23. Gagel J. J., Biemann K.: Anal. Chem. 58, 2184 (1986).
24. Sabo M., Gross J., Wang J. S., Rosenberg I. E.: Anal. Chem. 57, 1822 (1985).
25. Knehl D., Griffiths P. R.: J. Chromatogr. 17, 471 (1979).
26. Fujimoto C., Jinno K., Hirota Y.: J. Chromatogr. 258, 81 (1983).
27. Gagel J. J., Biemann K.: Anal. Chem. 59, 1266 (1987).

Index

Active centers on the surface,
132
Adrenaline, 105
Adsorbents, 116
Aerosol detector, 159
Amino acids, 170
3-Aminobenzoic acid, 172
4-Aminobenzoic acid, 173
4-Aminosalicylic acid, 173
Amount:
minimum detectable, 68, 74,
154
of the solute injected on the
column, 27
Amperometric detector, 75
signal, 64
Ampholyte, 172
Animals, tested, 169

Apiezone L, 135
Azo-t-butane, 150

1,2-Benzopyrene, 102
Buffer, 172

Capacity:
factor, 27
ratio, 13
column diameter dependence,
32
Capillary column, 124
flow rates, 124
inner diameter, 14, 129
open tubular, 15
Capillary columns:
fused silica, 127

[Capillary columns]
 liquid-adsorbent system, 133
 one-step-method preparation,
  126
 preparation, general, 125
 two-step-method preparation,
  126
Catecholamines, 169
Cation, competitive, 179
Cell:
 analysis, 166
 volume, 71
Cells, type of spectrometric
  detector, 66
4-Chlorphenol, 101, 102
Chromatogram, characteristics, 12
Column, 45
 capillary, development, 19
 device for packing, 46
 diameter, 13
 fused silica, 49
 material, 46
 micropacked, development, 18
 packed capillary, 15
 reduced length, 13, 156
 resistance, 129
 suppressor, 78
 thick-walled, 48
Complexes, 178
Concentration:
 analysis, 1
 counterion, 107
 limit, 78
 minimum detectable, 66, 77,
  79, 154, 158, 175, 179,
  180, 183

[Concentration]
 in the outlet from the column,
  95
 in the peak maximum, 14, 27,
  95
 of sample, 28
 solute in the sample, 14
 trace analysis, 3
 of traces, 94
Counterion, characteristics, 105,
  107
β-Cyclodextrin, 174

Dead time, 13
Dead volume, 13
Desorption, 118
Detection limits, 102, 167, 170
Detector:
 aerosol, 159
 amperometric, 158
 cell:
  IR, 196
  volume, 63, 153, 154
 concentration, 62
 conductometric, 78
 electrochemical, 75, 174
 flame ionization, 159
  with alkali metal, 160
 fluorimetric, 69
  mass, 64
 potentiometric, 158
 refractometric, 73
 time constant, 156, 158
Dialysis canula, 168
Diameter:
 column, reduced, 15

[Diameter]
   inner, of the capillary column, 14
   sorbent particle, 14, 37, 276
cis-Diaminodichlorplatinum, 174
1,2-Dibrom-2,2-
     dichloroethyldimethyl-
     phosphate, 143
2,4-Dichlorphenol, 101, 102
Dicumylperoxide, 150
Diffusion coefficient, 150
3,4-Dihydroxyphenylacetic acid,
   167
3,4-Dihydrohyphenylalanine, 167
5-Dimethylaminonaphtalene-1-
     sulfonylamino acids, 170
Dimethyloctadecylsilane, 167
Dimethylocytlamine, 167
Dispersion, 17
   extracolumn, 52
Distribution constant, 133
Dithiocarbamate, 179
Dopamine, 105, 167

Electrospray, 190
Elution strength, 17
Enrichment, 94
   column, 114
   technique, 96
Epinephrine, 105, 111
Etching, glass capillary, 130
Ethylenediamine, 182
Extracolumn dispersion, 52
Extracolumnar contributions, 130

Flame ionization detector, 159
   with alkali metal, 160

Flow rate, 14

Gradient:
   formation devices, 41
     categories, 41
     high-pressure part of a single
     pump, 42
   injection-generated, 17
   mobile-phase strength, 94

Height equivalent:
   of the reduced plate, 130
   of the theoretical plate, 14
   function of the mobile phase
     velocity, 53
   minimum, 54
   reduced, 13
Hydroquinone, 167
4-Hydroxybenzoic acid, 177
Hydroxycarboxylic acid, 179
5-Hydroxyindolacetic acid, 167
5-Hydroxytryptophan, 167

Immobilization of the stationary
   phase, 150
Infrared spectroscopy, 195
Inhomogeneity stationary phase,
   142
Injection:
   generated gradient, 17, 103
   stop-flow, 39
Interface LC/MS, 190
Ion chromatography, 78
Ion-exchange capacity, 151
Ionization sources, 189
Ion-pair chromatography, 103

Ions, inorganic, 178
Isotherm, 103, 107

Label, 170
Laser, 74
Layer, silica gel, 123

Mass analysis, 1
Mass spectrometry, 188
Mass trace analysis, 3
Mass of traces, 94
Mass transfer coefficient C, 55
Matrix, 6, 98
  elution strength, 36
20-Methylcholanthrene, 102
Microcolumn:
  chromatographs, producers, 86,
    88
  definition, 15
Miniaturization conditions, 9
Minimum detectable:
  amount, 154
  concentration, 154, 158, 175
  concentrations, 179, 180,
    183
Mixer:
  dynamic, 41
  static, 41
Mobile phase:
  gradient, 125
  swelling, 151
Modifier, 17
  concentration, 96

Naphthalene-2,3-dicarboxaldehyde,
  170

Neurotransmitters, 168
Nicotinic acid, 173
4-Nitro-2-aminophenol, 173
Noise, 64, 75
  fluorescent detector, 70
Noradrenaline, 105
Norepinephrine, 105, 111
Number of theoretical plates, 5,
  13
  column diameter dependence,
    32
  required, 31, 32

On-column ion chromatography,
  79
Optical fibers, 68, 72
Optical path, 66, 154

Packing device, capillary, 131
Particle, sorbent diameter, 37
Peak:
  concentration in maximum, 14,
    27, 95
  focusing, 17
    technique, 97
Perylene, 102
Phase ratio, 134, 150
pH-generated gradient, 172
Phenothiazine, 176
Photometric ultraviolet detector,
  65
Porosity interstitial, 13
Precolumns, 114
Pressure:
  drop in column, 29
  resistance, 123

Pump:
  constant flow, 38
  constant pressure, 38
    regime, 124
  efficiency, 40
  pneumatic, 39
  reciprocal, 39
  syringe, 39
  types, 39

Racemate, 174
Recovery, 168
Reduced quantities, 156
Refractometric detector, 73
Relative retention, 133, 151
Resolution, 32
Response:
  detectable, 6
  flame ionization detector with
    alkali metal, 161
Retention:
  relative, 32
  strength, 17, 109
  volume, 5, 13, 31

Sample:
  concentration, 10, 28
  enrichment, 17, 94
  injection, 79
    dispersion, 80, 82, 84
  matrix, 98
    elution strength, 36
  minimum detectable amount, 68
  volume, 28, 95, 100
    maximum applicable, 35
Secondary flow, 136

Separation principle of, 4
Serotonin, 167
Signal:
  amperometric detector, 64
  fluorescent detector, 70
  spectrophotometric detector, 63
Silicone phases, 135
Sorption system, 115
Spectrophotometric detector:
  signal, 63
  splitter, 158
Splitter, 41, 82, 124
Splitting ratio, 125
Standard deviation, 5
Stationary phase dynamically
  modified, 152
Strength:
  elution, 17
  retention, 17
Sulfacetamide, 175
Sulfadimidine, 175
Sulfanilamide, 175
Sulfanilic acid, 175
Sulfathiazole, 175
Surface:
  acid groups, 132
  active centres, 132
  cross-linked polymer, 133
  tension, 132, 133
Swelling of the mobile phase, 151

Tartaric acid, 182
Tetracyclines, 175
Theoretical plate, 5, 13, 14, 31
  reduced height equivalent, 27,
    130

Thioxanthene, 176
Trace analysis, nomenclature, 2
Trace concentration, 94
Trace mass, 94
Transport:
  detectors, 159
  system:
    IR, 196
    MS, 190
2,4,6-Trichlorphenol, 101, 102
Trimethylester phosphoric acid,
    143
1,2,3-Tris-2-cyanoethoxy-propane,
    135
Tryptophan, 167
Tyrosine, 167

Valve:
  four-way, 80
  six-way, 80
Variance, 13
Velocity:
  of the mobile phase, 27
    linear, 14
    optimal, 28, 54
Voltammogram, 167
Volume:
  detector cell, 63, 153, 154
  flow rate, 29
  of mobile phase, 13
  of stationary phase, 13
  variance of column, 51

9 780824 786410